Lecture Notes in Control and Information Sciences 209

Editor: M. Thoma

Springer-Verlag London Ltd.

Ciprian Foias, Hitay Özbay and Allen Tannenbaum

Robust Control of Infinite Dimensional Systems

Frequency Domain Methods

Springer

Authors

Ciprian Foias, Professor
Department of Mathematics, Indiana University, Bloomington, Indiana 47405, USA

Hitay Özbay, Associate Professor
Department of Electrical Engineering, The Ohio State University, Columbus, Ohio 43210, USA

Allen Tannenbaum, Professor
Department of Electrical Engineering, University of Minnesota, Minneapolis, Minnesota 55455, USA

ISBN 978-3-540-19994-6 ISBN 978-3-540-39386-3 (eBook)
DOI 10.1007/978-3-540-39386-3

British Library Cataloguing in Publication Data
Foias, Ciprian
Robust Control of Infinite Dimensional Systems: Frequency Domain Methods. - (Lecture Notes in Control & Information Sciences; Vol. 209)
I. Title II. Series
629.8312

Library of Congress Cataloging-in-Publication Data
A catalog record for this book is available from the Library of Congress

Originally published by Springer-Verlag London Limited in 1996.

Typesetting: Camera ready by authors

69/3830-543210 Printed on acid-free paper

THIS BOOK IS DEDICATED TO

Nicoleta, Dara, and Anta *C. Foias*

the memory of my grandfather *H. Özbay*

Rina, Emmanuel, and Sarah *A. Tannenbaum*

Preface

This aim of this book is to present a comprehensive treatment of $\mathcal{H}^\infty$ optimization techniques for linear time-invariant distributed parameter systems, e.g., systems with delays or those modelled by partial differential equations. We work strictly in the frequency domain which seems to us the natural context for the analysis and controller synthesis for such infinite dimensional plants. The underlying mathematical framework of the book is based on interpolation and dilation theory. All the relevant details of this subject will be presented here.

The prerequisites for this book are a working knowledge of classical $\mathcal{H}^\infty$ control, basic complex variables, elementary Hilbert space theory, and some working knowledge of partial differential equations. The material presented in this book have been used to teach a one-semester second year graduate course in Electrical Engineering. The book should be of interest to both theoretical engineers and mathematicians working in feedback control theory.

The project of considering $\mathcal{H}^\infty$ optimization of distributed parameter systems began in 1985 in our collaboration with Professor George Zames of McGill University. In fact, it was George Zames who suggested the whole project of considering sensitivity minimization of a plant consisting of a pure delay. We would like to thank George for all that he has taught us about the subject and for the insights which he has given us. Without him, this book may have never come into being. We would also like to thank the generous support given to us by the NSF, AFOSR, NASA and ARO during the writing of the text.

Contents

Chapter 1

Introduction

1.1 About the book

This book deals with certain robust control problems for a linear time invariant (LTI) infinite dimensional systems. Robust stabilization and sensitivity minimization problems (as well as disturbance attenuation in the sense of reducing the worst energy amplification from disturbance to an output signal) are studied in the framework of $\mathcal{H}^\infty$ control. In this setting the plant uncertainty is assumed to be dynamic (it has a transfer function). The book also includes a discussion on robust stabilization (stability margin optimization) under a parametric uncertainty. But the main focus is on the $\mathcal{H}^\infty$ control problem. An operator theoretic approach to this problem is presented here. This method, known as the *skew Toeplitz* theory, has been developed over the past few years (1987–1994), for several different cases: one, two and four block $\mathcal{H}^\infty$ optimal and suboptimal problems, stable and unstable plants, SISO (single input single output) and MIMO (multi input multi output) plants, etc. See the papers by H. Bercovici, C. Foias, A. Frazho, C. Gu, H. Özbay, M. C. Smith, A. Tannenbaum, O. Toker and G. Zames, [6], [28], [30], [33], [34], [35], [37], [38], [51], [79], [81], [82], [99], [101], etc. This book is based on these papers. The skew Toeplitz techniques have been applied to two benchmark problems: an unstable system with a time delay, and a flexible beam. These examples are described in our joint papers [20],

[67], [100] with D. Enns, K. Lenz, B. Morton, O. Toker and J. Turi. The flexible beam example is studied in Chapter 7. Several different time delay system examples appear in Chapters 4, 5, 6, and 7.

There are many other articles published on the $\mathcal{H}^\infty$ control of infinite dimensional systems. For example, the one block problem has been studied in [25], [48], [65], [64], [90], [114], [122]. Robust stabilization problem, for coprime factor perturbations of infinite dimensional systems, is a two block $\mathcal{H}^\infty$ problem and has been addressed in [42], [43], [71], [84], [112]. More general forms of $\mathcal{H}^\infty$ control problem for distributed parameter systems have been considered in [13], [27], [120], [92]. This list is not intended to be a complete literature survey on the $\mathcal{H}^\infty$ control of infinite dimensional systems. For a survey on this subject, see [14]. Most of the above mentioned papers approach the $\mathcal{H}^\infty$ control problem from the input/output operator theoretic point of view. The state space approach is more popular for the $\mathcal{H}^\infty$ control of finite dimensional systems. Because, in this case one can solve the problem from algebraic Riccati equations, which involve simple linear algebra, see e.g. [3], [17], [39], [47]. There are also game theoretic interpretations of these solutions, see e.g. [5] and references therein. Although it is possible to extend these results to infinite dimensional systems, in this case one has to be careful in using state space methods since more complicated semigroup theory and operator valued Riccati equations are involved. See [13] and [106] for the details of the state space $\mathcal{H}^\infty$ control problems for infinite dimensional systems. For a class of delay systems, numerical solutions to $\mathcal{H}^\infty$ control problems can also be obtained from finite dimensional Riccati equations, see e.g. [72] [95] and references therein.

In this book we shall see that, under certain mild assumptions, one can obtain $\mathcal{H}^\infty$ controllers (optimal and suboptimal), for distributed parameter systems, by solving a *set of finitely many linear equations*, which is called the *singular system.* Our purpose is to present basic steps of the skew Topelitz theory leading to these equations. We would like to emphesize that these finitely many equations are derived directly from the original infinite dimensional plant, i.e., no approximation is made. Recently H. Tu has developed a MATLAB program, [103], which constructs and solves these equations. Together with the formulae given in [80], this program computes the $\mathcal{H}^\infty$ optimal controllers for a class of distributed plants. For the same class of systems, O. Toker has obtained a much simplified version of the singular system equations, [98]–[101] to compute all suboptimal $\mathcal{H}^\infty$ controllers. A MATLAB based program can be obtained via e-mail by sending a request to H. Özbay.

The book is organized as follows. In Chapter 2 we give a mathematical background on linear operator theory and interpolation theory. Chapter 3 sets-up the $\mathcal{H}^\infty$ control problems related to robust stability and sensitivity minimization. Generalized stability margin optimization problem is also defined in this chapter. In Chapter 4, Nevanlinna-Pick interpolation approach to stability margin optimization, and optimal robustness/sensitivity problems, is presented. Also in this chapter, an operator theoretic approach is given for the solution of the standard one block $\mathcal{H}^\infty$ problem for stable distributed plants. In Chapter 5 we present generalizations of this solution to two block problem for unstable plants. Computation of suboptimal $\mathcal{H}^\infty$ controllers are discussed in Chapter 6. Two benchmark examples are given in Chapter 7. The status of the skew Toeplitz theory for the multivariable systems is discussed in Chapter 8, with the commutant lifting theorem. Finally in Chapter 9 we make some concluding remarks.

We have tried to keep the prerequisites to a minimum in writing this book to make it accessible to the widest possible control audience. We also tried to make the book accessible to mathematics audience, who may want to overlook certain explanatory paragraphs aimed at engineers. Basically what is needed is a good background in systems and some knowledge of $\mathcal{H}^\infty$ theory, say from [16]. We have tried to fill in most of the relevant mathematical details in order to make the book as self-contained as possible. However, courses in real and complex analysis will be very helpful to one's understanding. Some results presented in this book are stated without proofs for which the reader is referred to papers where they originally appear. The material of the book has already been course tested for second year control students at The Ohio State University and the University of Minnesota. The authors would like to thank Mr. Xing Guo, Dr. Thaddeus E. Peery and Dr. Onur Toker for carefully reading parts of the manuscript.

1.2 $\mathcal{H}^\infty$ control of distributed plants

The main reason to use feedback in the control of dynamical systems is to design against uncertainties. In a typical control system there are two kinds of uncertainties: modeling errors and disturbances. The purpose of feedback control is to achieve certain performance specifications in the closed loop system despite these uncertainties.

In control system design we start with a mathematical model of a given physical system. Infinite dimensional system models appear in many engineering applications where the physical system is spatially distributed, or contains time delays. For example, distillation columns [70], flexible beams [7], [68], heat conduction systems [105], aeroelastic systems [83] etc., can be cited as such engineering applications. For spatially distributed systems and systems with time delays partial differential equations or functional differential equations are taken as infinite dimensional mathematical models because these are the *simplest and most natural* representations of such systems, which give good physical insight. So, one of the reasons to use distributed models in the controller design is that infinite dimensional models may be more accurate in representing the dynamics of a physical system compared to finite dimensional models. On the other hand, in some cases infinite dimensional models which contain a few parameters are used for physical phenomenon which can otherwise be better explained by very high order finite dimensional models. Thus, the economical representation of the system is another important reason why distributed models are used in practice. For example there is only one parameter, h, in the representation of the time delay element e^{-hs}, which can be seen as an approximation to a finite dimensional model with many right half plane zeros, [20]. In general, transfer functions of distributed parameter systems are transcendental functions in the Laplace transform variable s, along with a few parameters, such as time delay, stiffness or damping coefficient of a beam.

In this book, the mathematical model describing a physical system will be assumed to be infinite dimensional. The controller has to be designed based on this *nominal plant model.* Since every model is an idealization of a much more complicated system, there is a modeling error. Of course it is impossible to characterize the error exactly (otherwise it would be possible to get an exact system description). On the other hand, it is possible to express modeling errors as perturbations of the nominal model, and most of the time it is possible to find an upper bound on these perturbations. Here we will consider perturbations of the nominal transfer function as modeling errors. Therefore, we are restricting ourselves to linear time invariant (LTI) perturbations of a LTI nominal model. Although this set-up ignores possible nonlinearities and time varying parameters in the actual system, it does handle an important class of modeling uncertainties. A nominal model transfer function $P(s)$, and a weighting function $W(s)$ (which represents an upper bound $|W(j\omega)|$ of the modeling error at each frequency $j\omega$), determine the "class of all possible plants," denoted by $\mathcal{P}$. We assume that the actual system, which is unknown, belongs to $\mathcal{P}$. Then, the robust stabilization problem is to find a fixed controller C, such that the closed loop system (represented by $[C, P_\Delta]$) is stable for all $P_\Delta \in \mathcal{P}$.

Besides stability, we should also study the effects of disturbances on the closed loop system behavior. We will assume that the disturbance is a finite energy signal. Then, the "effect" of the disturbance can be defined as the ratio of the output energy to the energy of the disturbance, i.e. energy amplification in the system. We can say that the closed loop system $[C, P]$ (resp. $[C, P_\Delta]$) has "good nominal (resp. robust) performance" if this energy amplification is "small" (resp. "small" for all $P_\Delta \in \mathcal{P}$).

In the text (see in particular Chapter 3), we will give definitions of robust stability, robust performance, and show that these problems can be put in the framework of the $\mathcal{H}^\infty$ control. This will allow us to formulate problems of robust system analysis and design in a *precise* mathematical manner which will make them amenable to techniques from operator theory and complex analysis. We will see that employing such a methodology, one can explicitly solve very general $\mathcal{H}^\infty$ problems in a simple, implementable manner. In fact, we will work our way up the hierarchy of control problems until we reach the general multivariable standard problem in Chapter 8.

Chapter 2

Mathematical Preliminaries

2.1 Notation

Below is the notation used throughout this book.

$\mathbb{Z}$: integers

$\mathbb{Z}_+$: non-negative integers, $\{n \in \mathbb{Z} \ : \ n \geq 0\} = \{0,\ 1,\ 2,\ \ldots\}$.

$\mathbb{R}$: real numbers.

$\mathbb{R}_+$: non-negative real numbers, $\{t \in \mathbb{R} \ : \ t \geq 0\} = [0\ ,\ \infty)$.

$\mathbb{C}$: complex numbers.

$\mathbb{C}_+$: open right half plane in $\mathbb{C}$, $\{s \in \mathbb{C} \ : \ \text{Re}\ s > 0\}$.

$\overline{\mathbb{C}_+}$: closed right half plane, $\{s \in \mathbb{C} \ : \ \text{Re}\ s \geq 0\}$.

$\widetilde{\mathbb{C}_+}$: extended right half plane, $\overline{\mathbb{C}_+} \cup \{\infty\}$.

$j\mathbb{R}$: imaginary axis, $\{s \in \mathbb{C} \ : \ \text{Re}\ s = 0\}$.

$j\mathbb{R}_e$: extended imaginary axis: $\{j\omega \ : \ \omega \in \mathbb{R} \cup \{\infty\}\}$.

$\mathbb{D}$: open unit disc, $\{z \in \mathbb{C} \ : \ |z| < 1\}$.

$\overline{\mathbf{D}}$: closed unit disc, $\{z \in \mathbf{C} \ : \ |z| \leq 1\}$.

$\mathbf{T}$: unit circle, $\{\zeta \in \mathbf{C} \ : \ |\zeta| = 1\}$.

$\overline{\alpha}$: complex conjugate of $\alpha \in \mathbf{C}$.

ess sup : essential supremum with respect to Lebesgue measure.

A^T : transpose of the matrix A.

A^* : transpose of the complex conjugate of A; when $\mathbf{A}$ is an operator, $\mathbf{A}^*$ denotes the adjoint of $\mathbf{A}$.

$\overline{\sigma}(A)$: largest singular value of A.

$\sigma(\mathbf{A})$: spectrum of the operator $\mathbf{A}$.

$\|\mathbf{A}\|$: norm of $\mathbf{A}$.

$\sigma_e(\mathbf{A})$: essential spectrum of $\mathbf{A}$.

$\|\mathbf{A}\|_e$: essential norm of $\mathbf{A}$.

$\mathcal{L}^1(\mathsf{R}_+)$: Lebesgue space of integrable real functions on R_+.

$\mathcal{L}^2(\mathsf{R}_+)$: Lebesgue space of square integrable real functions on R_+.

$\mathcal{L}^\infty(\mathsf{R}_+)$: Lebesgue space of essentially bounded real functions on R_+.

ℓ^1 : Real valued absolutely summable sequences on $\mathbf{Z}_+$.

ℓ^2 : Real valued square summable sequences on $\mathbf{Z}$.

ℓ^2_+ : Real valued square summable sequences on $\mathbf{Z}_+$.

ℓ^∞ : Real valued bounded sequences on $\mathbf{Z}_+$.

$\mathcal{L}^\infty(j\mathsf{R})$: Lebesgue space of essentially bounded functions on $j\mathsf{R}$.

$\mathcal{H}^\infty(\mathbf{C}_+)$: Hardy space of $\mathcal{L}^\infty(j\mathsf{R})$ functions which admit bounded analytical extensions to $\mathbf{C}_+$.

$\mathcal{L}^2(j\mathsf{R})$: Lebesgue space of square integrable functions on $j\mathsf{R}$.

$\mathcal{H}^2(\mathbb{C}_+)$: Hardy space of $\mathcal{L}^2(j\mathsf{R})$ functions which admit analytical extensions to $\mathbb{C}_+$.

$\mathcal{H}^1(\mathbb{C}_+)$: Hardy space of absolutely integrable functions on $j\mathsf{R}$ which admit analytical extensions to $\mathbb{C}_+$.

$\mathcal{L}^\infty(\mathsf{T}), \mathcal{H}^\infty(\mathsf{D}), \mathcal{L}^2(\mathsf{T}), \mathcal{H}^2(\mathsf{D}), \mathcal{H}^1(\mathsf{D})$: replace $j\mathsf{R}$ with T, and $\mathbb{C}_+$ with D in the above definitions of $\mathcal{L}^\infty(j\mathsf{R})$, $\mathcal{H}^\infty(\mathbb{C}_+)$, $\mathcal{L}^2(j\mathsf{R})$, $\mathcal{H}^2(\mathbb{C}_+)$, and $\mathcal{H}^1(\mathbb{C}_+)$, respectively.

$\mathcal{H}^\infty_{m\times n}(\mathbb{C}_+), \mathcal{H}^\infty_{m\times n}(\mathsf{D}), \mathcal{L}^\infty_{m\times n}(j\mathsf{R}), \mathcal{L}^\infty_{m\times n}(\mathsf{T})$: $m \times n$ matrix valued functions whose entries belong to $\mathcal{H}^\infty(\mathbb{C}_+)$, $\mathcal{H}^\infty(\mathsf{D})$, $\mathcal{L}^\infty(j\mathsf{R})$, $\mathcal{L}^\infty(\mathsf{T})$ respectively.

$\mathcal{H}^2_n(\mathbb{C}_+), \mathcal{H}^2_n(\mathsf{D}), \mathcal{L}^2_n(j\mathsf{R}), \mathcal{L}^2_n(\mathsf{T})$: $n \times 1$ vector valued functions with entries in $\mathcal{H}^2(\mathbb{C}_+)$, $\mathcal{H}^2(\mathsf{D})$, $\mathcal{L}^2(j\mathsf{R})$, $\mathcal{L}^2(\mathsf{T})$ respectively.

$\|G\|_p$: p-norm of G, when G is in $\mathcal{L}^p(j\mathsf{R})$, $\mathcal{H}^p(\mathbb{C}_+)$, $\mathcal{L}^p(\mathsf{T})$, $\mathcal{H}^p(\mathsf{D})$, etc.

$\mathcal{H}_1 \ominus \mathcal{H}_2$: orthogonal complement of $\mathcal{H}_2$ in $\mathcal{H}_1$, (where $\mathcal{H}_2$ is a subspace of a Hilbert space $\mathcal{H}_1$).

$\mathcal{K}^2_n$: $\mathcal{L}^2_n \ominus \mathcal{H}^2_n$ (could be on T or $j\mathsf{R}$).

$\mathbf{I}$: identity operator.

$\mathbf{P}_-$: orthogonal projection operator from $\mathcal{L}^2$ to $\mathcal{K}^2$

$\mathbf{P}_+ := \mathbf{I} - \mathbf{P}_-$.

$m\mathcal{H}^2(\mathsf{D}) := \{mf \ : \ f \in \mathcal{H}^2(\mathsf{D})\}$ where m is an inner function.

$\mathcal{H}(m) := \mathcal{H}^2(\mathsf{D}) \ominus m\mathcal{H}^2(\mathsf{D})$.

$\mathbf{P}_{\mathcal{H}}$: orthogonal projection onto a subspace $\mathcal{H}$ of $\mathcal{L}^2$ (on T or $j\mathsf{R}$).

2.2 Hardy spaces

A function $G(s)$, $s \in \mathbb{C}$, is in $\mathcal{H}^p(\mathbb{C}_+)$, $1 \leq p \leq \infty$, if

(i): G is analytic in $\mathbb{C}_+$,

(ii): it is defined almost everywhere on $j\mathsf{R}$, and

(iii): its p–norm defined by

$$\begin{aligned} \|G\|_p &= \sup_{\sigma>0} \left(\frac{1}{2\pi} \int_{-\infty}^{+\infty} |G(\sigma + j\omega)|^p d\omega \right)^{1/p}, \quad (1 \leq p < \infty) \\ &= (\text{ess} \sup_{\sigma>0,\ \omega\in\mathsf{R}} |G(\sigma + j\omega)|) \quad (p = \infty) \end{aligned}$$

is finite.

If G does not satisfy (i) but satisfies (ii) and (iii) with $\sigma = 0$, then it is in $\mathcal{L}^p(j\mathsf{R})$.

Similarly, a function $g(z)$, $z \in \mathbb{C}$, is in $\mathcal{H}^p(\mathsf{D})$, $1 \leq p \leq \infty$, if

(i): g is analytic in D,

(ii): it is defined almost everywhere on T, and

(iii): its p–norm defined by

$$\begin{aligned} \|g\|_p &= \sup_{r<1} \left(\frac{1}{2\pi} \int_{0}^{2\pi} |g(re^{j\theta})|^p d\theta \right)^{1/p}, \quad (1 \leq p < \infty), \\ &= (\text{ess} \sup_{r<1,\ \theta\in[0,2\pi]} |g(re^{j\theta})|) \quad (p = \infty) \end{aligned}$$

is finite.

If g does not satisfy (i) but satisfies (ii) and (iii) with $r = 1$, then it is in $\mathcal{L}^p(\mathsf{T})$.

The spaces $\mathcal{L}_n^2(j\mathsf{R})$ and $\mathcal{H}_n^2(\mathbb{C}_+)$ (resp. $\mathcal{L}_n^2(\mathsf{T})$ and $\mathcal{H}_n^2(\mathsf{D})$) $n \geq 1$, are Hilbert spaces, with the inner product

$$\begin{aligned} \langle G, F \rangle &:= \frac{1}{2\pi} \int_{-\infty}^{\infty} F(j\omega)^* G(j\omega) d\omega, \quad (\text{resp.} \\ \langle g, f \rangle &:= \frac{1}{2\pi} \int_{0}^{2\pi} f(e^{j\theta})^* g(e^{j\theta}) d\theta \ ). \end{aligned}$$

Note that the Laplace transform of $\mathcal{L}^2(\mathsf{R}_+)$ is the Hardy space $\mathcal{H}^2(\mathbb{C}_+)$. One can also see $\mathcal{L}^2_n(\mathsf{T})$ as the discrete Fourier transforms of ℓ^2 sequences, e.g. $g \in \mathcal{L}^2_n(\mathsf{T})$ has an expansion

$$g(e^{j\theta}) = \sum_{k=-\infty}^{\infty} g_k e^{jk\theta},$$

with $g_k \in \mathbb{C}^n$ and

$$\|g\|_2^2 = \sum_{k=-\infty}^{\infty} g_k^T g_k < \infty.$$

The second Hardy space $\mathcal{H}^2_n(\mathsf{D})$, (a subspace of $\mathcal{L}^2_n(\mathsf{T})$), is the space of discrete Fourier transforms of ℓ^2_+ sequences, i.e. $g \in \mathcal{H}^2_n(\mathsf{D})$ if and only if $g \in \mathcal{L}^2_n(\mathsf{T})$ and $g_k = 0$ for $k < 0$; in this case the Z-transform $g(z) = \sum_{k=0}^{\infty} g_k z^k$ converges for all $z \in \mathsf{D}$.

It is also important to note that

$$\mathcal{H}^\infty(\mathsf{D}) = \mathcal{L}^\infty(\mathsf{T}) \cap \mathcal{H}^2(\mathsf{D}).$$

If $G \in \mathcal{L}^\infty_{m\times n}(j\mathsf{R})$ (resp. $g \in \mathcal{L}^\infty_{m\times n}(\mathsf{T})$) then its ∞-norm is defined as

$$\|G\|_\infty = \text{ess} \sup_{\omega\in\mathsf{R}} \overline{\sigma}(G(j\omega)) \quad \text{(resp. } \|g\|_\infty = \text{ess} \sup_{\theta\in[0,2\pi]} \overline{\sigma}(g(e^{j\theta}))\text{)}.$$

The 2-norm of a vector valued function $G \in \mathcal{L}^2_n(j\mathsf{R})$ (resp. $g \in \mathcal{L}^2_n(\mathsf{T})$) is defined as

$$\begin{aligned} \|G\|_2 &= \left(\frac{1}{2\pi}\int_{-j\infty}^{+j\infty} G(j\omega)^* G(j\omega) d\omega\right)^{1/2}, \quad \text{(resp.} \\ \|g\|_2 &= \left(\frac{1}{2\pi}\int_0^{2\pi} g(e^{j\theta})^* g(e^{j\theta}) d\theta\right)^{1/2} \quad). \end{aligned}$$

Any $n \times n$ matrix U whose entries are in $\mathcal{L}^\infty(j\mathsf{R})$ (or in $\mathcal{L}^\infty(\mathsf{T})$) with the property

$$\begin{aligned} U(j\omega)^* U(j\omega) = U(j\omega) U(j\omega)^* &= \mathrm{I}_{n\times n} \text{ a.e. } \omega \in \mathsf{R} \text{ (or} \\ U(e^{j\theta})^* U(e^{j\theta}) = U(e^{j\theta}) U(e^{j\theta})^* &= \mathrm{I}_{n\times n} \text{ a.e. } \theta \in [0, 2\pi] \text{)} \end{aligned}$$

is called ***unitary.*** Unitary matrices preserve the norm, i.e. if U is an $n \times n$ matrix valued function which is unitary then we have $\|UL\|_\infty = \|U^*L\|_\infty = \|L\|_\infty$ for all L in $\mathcal{L}^\infty_{n\times m}$ (of $j\mathbb{R}$ or $\mathbb{T}$). For such U, we also have $\|UL\|_2 = \|U^*L\|_2 = \|L\|_2$, for all L in $\mathcal{L}^2_n$ (of $j\mathbb{R}$ or $\mathbb{T}$).

In the rest of this book when we refer to Lebesgue or Hardy spaces on the imaginary axis, right half plane, unit circle, or unit disc we will drop the arguments ($j\mathbb{R}$), ($\mathbb{C}_+$), ($\mathbb{T}$), and ($\mathbb{D}$) whenever the meaning is clear from the context.

2.3 Conformal map between $\mathbb{C}_+$ and $\mathbb{D}$

In this book, the systems are represented by their transfer functions, which are functions of the Laplace transform variable $s \in \mathbb{C}_+$ (in the case of continuous time systems) or functions of the Z-transform variable $z \in \mathbb{D}$ (for discrete time systems). Our solution to the $\mathcal{H}^\infty$ control problems will be derived using functions defined on the unit disc ($z-$plane). This does not limit us to discrete time systems, because we can transform a continuous time problem to a discrete time problem via a conformal map between $\mathbb{C}_+$ and $\mathbb{D}$. A simple example of such a map is

$$z = \frac{s-a}{s+a}, \quad s = a\frac{1+z}{1-z}, \quad a > 0$$

where $s \in \mathbb{C}_+$ and $z \in \mathbb{D}$. This conformal map transforms every point in $\mathbb{C}_+$ to a unique point in $\mathbb{D}$ and vice versa, the imaginary axis (boundary of $\mathbb{C}_+$) is mapped to the unit circle (boundary of $\mathbb{D}$). In particular, the points $j\infty$ and 0 in the $s-$plane are mapped to the points 1 and -1 in the $z-$plane.

Any function $F \in \mathcal{H}^\infty$ defined on $\mathbb{C}_+$ can be represented in terms of a function $f \in \mathcal{H}^\infty(\mathbb{D})$, and vice versa, e.g. choosing $a = 1$:

$$f(z) = F(\frac{1+z}{1-z}) \quad \text{and} \quad F(s) = f(\frac{s-1}{s+1}).$$

The conformal map between $\mathbb{C}_+$ and $\mathbb{D}$ preserves all the important properties of $F(s)$ as a bounded analytic function, e.g., $f(z)$ is a bounded analytic function on $\mathbb{D}$ and

$$\|F\|_\infty = \text{ess} \sup_{\omega \in \mathbb{R}} |F(j\omega)| = \text{ess} \sup_{\theta \in [0,2\pi]} |f(e^{j\theta})| = \|f\|_\infty .$$

In view of the above remarks, we can transform the problem data from $\mathbb{C}_+$ to $\mathbb{D}$. For example, choosing $a = 1$, if $P(s)$ represents the transfer function of the plant, then it can also be represented by $p(z) = P(\frac{1+z}{1-z})$, as a function defined on the unit disc. Conversely, if the controller is given as a function of z, i.e. $c(z)$, then, its transfer function can be obtained from the inverse map, i.e. $C(s) = c(\frac{s-1}{s+1})$.

2.4 Bounded linear operators

2.4.1 Operator norm and the essential norm

Consider two Banach spaces $\mathcal{K}_1$ and $\mathcal{K}_2$, with norms denoted by $\|\cdot\|_{\mathcal{K}_1}$ and $\|\cdot\|_{\mathcal{K}_2}$, respectively. Let $\mathbf{L}$ be a linear operator from $\mathcal{K}_1$ to $\mathcal{K}_2$. Then, $\mathbf{L}$ is bounded if its norm, defined by

$$\|\mathbf{L}\| := \sup \left\{ \frac{\|\mathbf{L}x\|_{\mathcal{K}_2}}{\|x\|_{\mathcal{K}_1}} \quad : \quad x \in \mathcal{K}_1 \ , \quad x \neq 0 \right\},$$

is finite.

Suppose $\mathcal{K}_1$ and $\mathcal{K}_2$ are two separable Hilbert spaces and $\mathbf{L}$ is a bounded linear operator from $\mathcal{K}_1$ to $\mathcal{K}_2$. Then the norm of $\mathbf{L}$ is given by

$$\|\mathbf{L}\| = \max\{\|\mathbf{L}\|_e \ , \ \sigma_{max}\},$$

where σ_{max} denotes the largest singular value of finite multiplicity, of $\mathbf{L}$, and $\|\mathbf{L}\|_e$ denotes the essential norm.

Recall that a singular value of an operator $\mathbf{L}$ is the positive square root of an eigenvalue of the operator $\mathbf{L}^*\mathbf{L}$; it is of finite multiplicity if the eigenvalue has finite multiplicity. To define the essential norm let $\langle\cdot,\cdot\rangle_1$, denote the inner product on $\mathcal{K}_1$. We say that a sequence $x_n \in \mathcal{K}_1$, $n = 1,2,3,\ldots$, *converges to zero weakly* if

$$\langle y, x_n\rangle_1 \to 0, \quad \text{as} \quad n \to \infty, \quad \text{for all} \quad y \in \mathcal{K}_1.$$

Then, the essential norm of $\mathbf{L}$ is given by

$$\|\mathbf{L}\|_e = \max\{\sqrt{\lambda} \quad : \quad \lambda \in \sigma_e(\mathbf{L}^*\mathbf{L})\},$$

where $\sigma_e(\mathbf{L}^*\mathbf{L})$ denotes the essential spectrum of $\mathbf{L}^*\mathbf{L}$ which consists of those $\lambda \in \mathbb{C}$, for which there exists a sequence $x_n \in \mathcal{K}_1$, with $\langle x_n \ , \ x_n\rangle_1 = 1$ for all $n = 1,2,\ldots$, and $x_n \to 0$ weakly as $n \to \infty$, such that

$$\|(\lambda\mathbf{I} - \mathbf{L}^*\mathbf{L})x_n\|_{\mathcal{K}_1} \to 0 \quad \text{as} \quad n \to \infty.$$

2.4.2 $\mathcal{H}^\infty_{m\times n}$ as bounded linear operators on $\mathcal{H}^2_n$

We can see the elements of $\mathcal{H}^\infty_{m\times n}$ as bounded linear multiplication operators on $\mathcal{H}^2_n$. More precisely, if $G \in \mathcal{H}^\infty_{m\times n}$, then it defines a multiplication operator $\mathbf{M}_G : \mathcal{H}^2_n \to \mathcal{H}^2_m$

$$\mathbf{M}_G f = Gf \ , \quad f \in \mathcal{H}^2_n, \tag{2.1}$$

with the following property.

Theorem 1 *Let G be a matrix valued function in $\mathcal{H}^\infty_{m\times n}$. Then, $\mathbf{M}_G f \in \mathcal{H}^2_m$ for all $f \in \mathcal{H}^2_n$, and*

$$\|\mathbf{M}_G\| = \|G\|_\infty,$$

where $\mathbf{M}_G$ is the operator defined by (2.1).

Proof. We will give the proof for $n = m = 1$; but it can be extended to the general case easily. First note that multiplications of analytic functions give rise to analytic functions. Also, since both G and f are defined a.e. on the boundary (i.e. $j\mathbb{R}$ or $\mathbb{T}$), Gf is also defined a.e. on the boundary. Therefore, we just need to show that

$$\|\mathbf{M}_G\| = \sup\left\{\frac{\|Gf\|_2}{\|f\|_2} \quad : \quad f \in \mathcal{H}^2 \, , \, f \neq 0\right\} = \|G\|_\infty .$$

Note that for any $f \in \mathcal{H}^2$ we have

$$\begin{aligned}
\|Gf\|_2^2 &= \frac{1}{2\pi}\int_{-\infty}^{\infty} |G(j\omega)f(j\omega)|^2 d\omega \\
&= \frac{1}{2\pi}\int_{-\infty}^{\infty} |G(j\omega)|^2 |f(j\omega)|^2 d\omega \\
&\leq \|G\|_\infty^2 \frac{1}{2\pi}\left(\int_{-\infty}^{\infty} |f(j\omega)|^2 d\omega\right) \\
&\leq \|G\|_\infty^2 \|f\|_2^2 .
\end{aligned}$$

Hence, $\|\mathbf{M}_G\| \leq \|G\|_\infty$. In order to see the converse recall the definition

$$\|G\|_\infty = \operatorname*{ess\,sup}_{\omega \in \mathbb{R}} |G(j\omega)| .$$

This means that for every $\epsilon > 0$ there exists a finite number $\delta > 0$ and a measurable set Ω of measure δ, such that

$$|G(j\omega)| \geq \|G\|_\infty - \epsilon \quad \text{for all} \quad \omega \in \Omega \, .$$

On the other hand we can find a function $f_o \in \mathcal{H}^2$ such that $\|f_o\|_2 = 1$ and

$$|f_o(j\omega)| \geq (1-\epsilon)\sqrt{\frac{2\pi}{\delta}} \quad \text{for} \quad \omega \in \Omega \, .$$

One can construct such a function as follows. Let $f(s)$ in $\mathcal{H}^2(\mathbb{C}_+)$ be defined by the integral

$$f(s) = exp(\frac{1}{\pi}\int_{-\infty}^{\infty}\left(\frac{j\omega s - 1}{s - j\omega}\right)\frac{\varphi(\omega)d\omega}{\omega^2 + 1})$$

where

$$\varphi(\omega) = -\ln|f(j\omega)| .$$

So, given a desired magnitude function $\varphi(\omega)$ on Ω, such an $f_o(s)$ can be constructed. Thus, the 2-norm of $y_o = Gf_o$ is bounded below by

$$\begin{aligned} \|y_o\|_2 &\geq \left(\frac{1}{2\pi}\int_\Omega (\|G\|_\infty - \epsilon)^2 (\frac{2\pi}{\delta})(1-\epsilon)^2 d\omega\right)^{\frac{1}{2}} \\ &\geq (\|G\|_\infty - \epsilon)(1-\epsilon). \end{aligned}$$

Since ϵ can be made arbitrarily small we have

$$\|\mathbf{M}_G\| \geq \|G\|_\infty .$$

This concludes the proof for $n = m = 1$ □.

The above proof can be extended to multivariable case as follows: Again, it is easy to establish the upper bound. For the lower bound same argument works except that now for a fixed matrix $G(j\omega_o)$ we choose a fixed singular vector $f_o(j\omega_o)$ such that

$$\|G(j\omega_o)f_o(j\omega_o)\|^2 = \overline{\sigma}(G(j\omega_o))^2\|f_o(j\omega_o)\|^2.$$

For the case where the functions are defined on the unit disc $\mathbf{D}$, the proof is still valid with obvious modifications.

2.5 The shift operator

The *shift operator*, is defined as $\mathbf{S} : \mathcal{H}^2(\mathbf{D}) \to \mathcal{H}^2(\mathbf{D})$

$$(\mathbf{S}f)(z) = zf(z) = 0 + f_0 z^1 + f_1 z^2 + \ldots,$$

for all $f \in \mathcal{H}^2(\mathbf{D})$, with $f(z)$ being the Z-transform of the ℓ^2_+ sequence $\{f_k\}_{k=0}^\infty$. So, $\mathbf{S}$ "shifts the coefficients to the right."

The adjoint of the shift operator, denoted by $\mathbf{S}^*$, "shifts the coefficients to the left" as follows, $\mathbf{S}^* : \mathcal{H}^2(\mathsf{D}) \to \mathcal{H}^2(\mathsf{D})$

$$(\mathbf{S}^* f)(z) = z^{-1}(f(z) - f_0) = f_1 + f_2 z^1 + f_3 z^2 + \dots$$

for all $f \in \mathcal{H}^2(\mathsf{D})$, as before.

An important point to remark is that $\mathbf{S}^{*k}\mathbf{S}^k$ is the identity (for any integer $k \geq 1$), however $\mathbf{S}^k\mathbf{S}^{*k} \neq \mathbf{I}$:

$$\begin{aligned} (\mathbf{S}^{*k}\mathbf{S}^k f)(z) &= f(z), \\ (\mathbf{S}^k\mathbf{S}^{*k} f)(z) &= f(z) - \sum_{\ell=0}^{k-1} f_\ell z^\ell = \sum_{\ell=k}^{\infty} f_\ell z^\ell. \end{aligned}$$

Note that $\mathbf{S}$ can be seen as the multiplication operator $\mathbf{M}_g$, where $g(z) := z$. Moreover, by Theorem 1 we have

$$\|\mathbf{S}\| = \|\mathbf{M}_g\| = \|g\|_\infty = \operatorname{ess} \sup_{\theta \in [0,2\pi]} |e^{j\theta}| = 1.$$

Conversely, we can define multiplication operators using the shift operator. For example given any $g \in \mathcal{H}^\infty(\mathsf{D})$ the operator $g(\mathbf{S})$ is obtained by formally replacing z by $\mathbf{S}$ in the power series expansion of $g(z)$:

$$g(\mathbf{S}) = \sum_{k=0}^{\infty} g_k \mathbf{S}^k \quad \text{and} \quad (g(\mathbf{S})f)(z) = g(z)f(z)$$

for all $f \in \mathcal{H}^2(\mathsf{D})$. Note that by definition $g(\mathbf{S}) = \mathbf{M}_g$.

2.6 Inner-Outer factorizations

Definition: A function $m \in \mathcal{H}^\infty(\mathsf{D})$ is called *inner* if $|m(z)| \leq 1$ for all $z \in \mathsf{D}$ and $|m(e^{j\theta})| = 1$ a.e. $\theta \in [0, 2\pi]$.

Since inner functions have constant magnitude a.e. on T; the engineers will realize that they generalize *all pass* transfer functions. An important property of inner functions is that they do not change the norm, i.e. both $m \in \mathcal{H}^\infty(\mathsf{D})$ and $m^* \in \mathcal{L}^\infty(\mathsf{T})$ are unitary.

Examples of inner functions include

$$\begin{aligned} m_1(z) &= \frac{z-a}{1-az}, \quad a \in (-1,1) \\ m_2(z) &= e^{-h\frac{1+z}{1-z}}, \quad h > 0 \\ m_3(z) &= \prod_{k=1}^{\infty}\left(\frac{z-a_k}{1-\overline{a_k}z}\right), \quad |a_k| < 1, \quad \text{and} \quad \sum_{k=1}^{\infty}(1-|a_k|) < \infty \\ m_4(z) &= m_1(z)m_2(z)m_3(z). \end{aligned}$$

In the above examples we see that $m_2(z)$ and $m_4(z)$ are not defined at $z = 1$. Also note that $a_\infty := \lim_{k\to\infty} a_k$ must lie on the unit circle, and at that point $m_3(z)$ and $m_4(z)$ are not well defined. Such points on the unit circle are the *essential singularities* of inner functions. We would like to point out that rational inner functions have no essential singularities.

Theorem 2 (Beurling) *Let $\mathcal{M}$ be a closed subspace of $\mathcal{H}^2(\mathsf{D})$ which is invariant with respect to* $\mathbf{S}$ *(i.e. $\mathbf{S}\mathcal{M} = \{\mathbf{S}f \; : \; f \in \mathcal{M}\}$ is a subspace of $\mathcal{M}$). Then, there exists an inner function $m \in \mathcal{H}^\infty(\mathsf{D})$ such that*

$$m\mathcal{H}^2(\mathsf{D}) = \mathcal{M}.$$

Conversely, given any inner function m, the subspace $\mathcal{M}_m := m\mathcal{H}^2(\mathsf{D})$ is closed in $\mathcal{H}^2(\mathsf{D})$ and invariant under the shift operator $\mathbf{S}$. □

This theorem characterizes the shift invariant subspaces of $\mathcal{H}^2(\mathsf{D})$ in terms of the inner functions.

Definition: A function $g \in \mathcal{H}^\infty(\mathsf{D})$ is called *outer* if the closure of $g\,\mathcal{L}_+$ in $\mathcal{H}^2(\mathsf{D})$ is the whole space $\mathcal{H}^2(\mathsf{D})$, where $\mathcal{L}_+ = \{\sum_{k=0}^{n} a_k z^k, \quad a_k \in \mathbb{C}, \quad n \geq 0\}$.

Outer functions generalize *minimum phase* functions: they don't have a zero in D, but may have zeros on T. So, if g is an outer function, with $\inf_\theta |g(e^{j\theta})| > 0$, then it is *invertible* in $\mathcal{H}^\infty(\mathbf{D})$, i.e. there is another outer function $h \in \mathcal{H}^\infty(\mathbf{D})$ such that $g(z)h(z) = 1$, for all $z \in \mathbf{D}$.

Theorem 3 *([94]) Let f be a function in $\mathcal{H}^\infty(\mathbf{D})$, then it admits an inner-outer factorization of the form*

$$f(z) = m(z)g(z),$$

where m is inner and g is outer. Note that $|f(e^{j\theta})| = |g(e^{j\theta})|$ a.e. $\theta \in [0, 2\pi]$ and hence $\|f\|_\infty = \|g\|_\infty$.□

An inner/outer factorization can be done by finding a spectral factor $g(e^{j\theta})$ of $|f(e^{j\theta})|^2$. Whenever $f(z)f(z^{-1})$ is a rational function, g is finite dimensional, and it can be found by solving an algebraic Riccati equation, see for example [39]. In the general case it might be difficult to find the inner/outer factorization. On the other hand, for several interesting situations where $f(z)f(z^{-1})$ is irrational, it is still possible to compute the inner/outer factorizations, see e.g. [67] for a flexible beam system example. Here we present a delay system example from [78].

Example: Let us consider

$$g(z) = \frac{e^{-h\frac{1+z}{1-z}}(\frac{1+z}{1-z} - 3)(\frac{1+z}{1-z} - 0.5)}{(\frac{1+z}{1-z} + 2)^2(\frac{1+z}{1-z} + 0.1 - e^{-h_1\frac{1+z}{1-z}})}, \quad h_1 = 2\ln(\frac{5}{3}), \quad h > 0.$$

Note that the only point in $\mathbf{D}$ where the term $(\frac{1+z}{1-z} + 0.1 - e^{-h_1\frac{1+z}{1-z}})$ becomes zero is $z = 1/3$. We can easily check that the multiplicity of this zero is 1. On the other hand the term $(\frac{1+z}{1-z} - 0.5)$, in the numerator, also becomes zero at $z = 1/3$. So, g is bounded in $\mathbf{D}$. The inner and outer parts of g are

$$\begin{aligned} m(z) &= e^{-h\frac{1+z}{1-z}}\left(\frac{z-0.5}{1-0.5z}\right) \\ f(z) &= g(z)/m(z). \end{aligned}$$

We conclude this section by noting that in control theory another classification of analytic functions is useful. A function $F(s)$ defined on $\mathbb{C}_+$ is called *proper* (resp. *strictly proper*) if

$$\lim_{|s|\to\infty} |F(s)| < \infty \quad (\text{resp.} \quad \lim_{|s|\to\infty} |F(s)| = 0).$$

Similar definitions can be made for functions defined on $\mathbf{D}$, with respect to a distinguished point in $\mathbf{T}$.

2.7 The compressed shift operator

In order to define the compressed shift operator, we will first need a few elementary results about Hilbert spaces. First let $\mathcal{H}$ denote an arbitrary (complex, separable) Hilbert space, and $\mathcal{H}_1 \subset \mathcal{H}$ a Hilbert subspace. We define

$$\mathcal{H} \ominus \mathcal{H}_1 := \{h \in \mathcal{H} : \langle h, h_1 \rangle = 0 \ \ \forall h_1 \in \mathcal{H}_1\}.$$

$\mathcal{H} \ominus \mathcal{H}_1$ is called the *orthogonal complement* of $\mathcal{H}_1$ in $\mathcal{H}$. One can show the following (see e.g. [54])

Theorem 4 *Let $h \in \mathcal{H}$. Then there exist unique vectors $h_1 \in \mathcal{H}_1$, $h_2 \in \mathcal{H} \ominus \mathcal{H}_1$, such that $h = h_1 + h_2$.*

Using the notation of Theorem 4, we define an operator $\mathbf{P}_1 : \mathcal{H} \to \mathcal{H} \ominus \mathcal{H}_1$ by setting $\mathbf{P}_1 h = h_2$ for each $h \in \mathcal{H}$. $\mathbf{P}_1$ is called the *orthogonal projection* of $\mathcal{H}$ onto $\mathcal{H} \ominus \mathcal{H}_1$.

Now given an inner function $m \in \mathcal{H}^\infty(\mathbf{D})$, by Theorem 2 (Beurling's theorem), $m\mathcal{H}^2(\mathbf{D}) \subset \mathcal{H}^2(\mathbf{D})$ is a closed shift-invariant subspace and every closed shift-invariant subspace of $\mathcal{H}^2(\mathbf{D})$ has this form, [94]. We can consider therefore the Hilbert space $\mathcal{H}(m) := \mathcal{H}^2(\mathbf{D}) \ominus m\mathcal{H}^2(\mathbf{D})$ and the corresponding orthogonal projection $\mathbf{P}_{\mathcal{H}(m)} : \mathcal{H}^2(\mathbf{D}) \to \mathcal{H}(m)$. So, any function $h \in \mathcal{H}^2(\mathbf{D})$ has an orthogonal decomposition $h = g + mf$ where $f \in \mathcal{H}^2(\mathbf{D})$ and $g \in \mathcal{H}(m)$. If $g \in \mathcal{H}(m)$ then m^*g is of the form

$$(m^*g)(\zeta) = \overline{m(\zeta)}g(\zeta) = \sum_{i=1}^{\infty} \phi_{-i}\zeta^{-i} \quad \text{for} \quad \zeta \in \mathbf{T} \tag{2.2}$$

where the right hand side converges a.e. on $\mathbf{T}$ and outside the unit disc, for some coefficients $\phi_{-i} \in \mathbf{C}, i \geq 1$ such that $\sum_{i=1}^{\infty} |\phi_{-i}|^2 < \infty$. In other words the function $g_\perp := m^*g$ is in $\mathcal{L}^2(\mathbf{T}) \ominus \mathcal{H}^2(\mathbf{D})$.

Before giving a precise definition of the compressed shift operator we would like to present some special properties of $\mathcal{H}(m)$ and $\mathbf{P}_{\mathcal{H}(m)}$ when m is rational. If m is a rational inner function then it is of the form $m = b_1 b_2$, where $b_1(z) = z^n$ and

$$b_2(z) = \prod_{k=1}^{\ell} \left(\frac{z - a_k}{1 - \overline{a_k} z} \right),$$

with $|a_k| < 1$, for $k = 1, \ldots, \ell$. When $m = b_1 b_2$, where b_1 and b_2 are as above, $\mathcal{H}(m) = \mathcal{H}(b_1 b_2)$ has an orthogonal decomposition of the form

$$\begin{aligned} \mathcal{H}(b_1 b_2) &= \mathcal{H}^2(\mathbf{D}) \ominus b_1 b_2 \mathcal{H}^2(\mathbf{D}) \\ &= (\mathcal{H}^2(\mathbf{D}) \ominus b_1 \mathcal{H}^2(\mathbf{D})) \oplus b_1(\mathcal{H}^2(\mathbf{D}) \ominus b_2 \mathcal{H}^2(\mathbf{D})) \\ &= \mathcal{H}(b_1) \oplus b_1 \mathcal{H}(b_2). \end{aligned}$$

Moreover, $\mathcal{H}(b_1)$ and $\mathcal{H}(b_2)$, (and hence $\mathcal{H}(b_1 b_2)$) are finite dimensional by the following results. For simplicity we will assume that $a_i \neq a_j$ for $i \neq j$, $1 \leq i, j \leq \ell$, i.e. the zeros of b_2 are distinct.

Lemma 1 *$\mathcal{H}(b_1)$ is a finite dimensional vector space of dimension n. A basis of $\mathcal{H}(b_1)$ consists of the elements $\{1,\ z, \ldots,\ z^{n-1}\}$.*

Proof. First note that the usual orthonormal basis for $\mathcal{H}^2(\mathbf{D})$ is given by $\{1,\ z,\ z^2, \ldots\}$. Then, in terms of these basis functions $b_1 \mathcal{H}^2(\mathbf{D})$ has the basis $\{z^n,\ z^{(n+1)},\ z^{(n+2)}, \ldots\}$, because $b_1(z) = z^n$. Hence, $\{1,\ z, \ldots,\ z^{n-1}\}$ is an orthonormal basis for $\mathcal{H}(b_1)$. □

Lemma 2 *Assume that the zeros of b_2 are distinct. Then, $\mathcal{H}(b_2)$ is a finite dimensional vector space of dimension ℓ; and a basis of $\mathcal{H}(b_2)$ consists of the elements $\{f_1, \ldots, f_\ell\}$ where*

$$f_i(z) := \frac{1}{1-\overline{a_i}z} \quad \text{for } i = 1, \ldots, \ell\,.$$

Proof. For any $q \in \mathcal{H}^2(\mathbf{D})$, $q \in b_2\mathcal{H}^2(\mathbf{D})$ if and only if $q(a_i) = 0$ for $i = 1, \ldots, \ell$. But by the Cauchy integral theorem

$$\langle h, f_i \rangle = \frac{1}{2\pi}\int_0^{2\pi} \frac{h(e^{j\theta})}{1-a_i e^{-j\theta}}d\theta = h(a_i), \quad \text{for } h \in \mathcal{H}^2(\mathbf{D}).$$

Thus, $q \in b_2\mathcal{H}^2(\mathbf{D})$ if and only if

$$\langle q, f_i \rangle = 0 \quad \text{for } i = 1, \ldots, \ell. \tag{2.3}$$

But (2.3) is equivalent to the condition that q is orthogonal to the linear span of the f_i, which we denote by $\mathcal{V}$. In other words $q \in b_2\mathcal{H}^2(\mathbf{D})$ if and only if $q \perp \mathcal{V}$, which means $\mathcal{V} = \mathcal{H}(b_2)$ as required.

We have shown that the f_i's span $\mathcal{H}(b_2)$. We must now show that they are linearly independent. Suppose to the contrary that they are linearly dependent. Then, we may clearly suppose (after perhaps renumbering the f_i's) that f_1 is in the linear span of $f_2, \ldots, f_\ell$. In this case, we would have that

$$\mathcal{H}^2(\mathbf{D}) \ominus b_2\mathcal{H}^2(\mathbf{D}) = \mathcal{H}^2(\mathbf{D}) \ominus \hat{b}_2\mathcal{H}^2(\mathbf{D})$$

where

$$\hat{b}_2(z) = \prod_{i=2}^{\ell}\left(\frac{z-a_i}{1-\overline{a_i}z}\right).$$

Thus $b_2\mathcal{H}^2(\mathbf{D}) = \hat{b}_2\mathcal{H}^2(\mathbf{D})$. This means that a function $q \in \mathcal{H}^2(\mathbf{D})$ vanishes at $a_1, \ldots, a_\ell$ if and only if it vanishes at $a_2, \ldots, a_\ell$ which is clearly absurd. □

In general when

$$b_2(z) = \prod_{i=1}^{n} \left(\frac{z - a_i}{1 - \bar{a}_i z} \right)^{n_i}$$

with $a_i \neq a_j$ for $i \neq j$, we define $k = \max_{1 \leq i \leq n} n_i$, and let

$$b_{21}(z) = \prod_{i=1}^{n} \left(\frac{z - a_i}{1 - \bar{a}_i z} \right) , \quad b_{22}(z) = \prod_{n_i \geq 2} \left(\frac{z - a_i}{1 - \bar{a}_i z} \right) , \cdots$$

$$b_{2k}(z) = \prod_{n_i \geq k} \left(\frac{z - a_i}{1 - \bar{a}_i z} \right) .$$

Then, as before, it is easy to show that $b_2 = b_{21} b_{22} \ldots b_{2k}$ and

$$\mathcal{H}(b_2) = \mathcal{H}(b_{21}) \oplus b_{21}\mathcal{H}(b_{22}) \oplus b_{21}b_{22}\mathcal{H}(b_{23}) \oplus \cdots \oplus b_{21} \ldots b_{2(k-1)}\mathcal{H}(b_{2k}).$$

Therefore,

$$\dim\mathcal{H}(b_2) = \sum_{i=1}^{n} \dim\mathcal{H}(b_{2i}) = \deg(b_2).$$

The above results show that whenever m is rational $\mathcal{H}(m)$ is a finite dimensional subspace of $\mathcal{H}^2(\mathbf{D})$. This observation becomes very important in the solution of $\mathcal{H}^\infty$ control problems associated with finite dimensional systems, or infinite dimensional plants with finitely many unstable modes. On the other hand, if m is irrational, (or an infinite Blaschke product), the space $\mathcal{H}(m)$ is infinite dimensional.

In the finite dimensional case, i.e. when m is rational, it is easy to see the action of $\mathbf{P}_{\mathcal{H}(m)}$ on an arbitrary element $h \in \mathcal{H}^2(\mathbf{D})$, because $\mathbf{P}_{\mathcal{H}(m)}$ is a finite rank operator. To illustrate this point let us consider $m = b_2$, defined above with distinct zeros. Then, by Lemma 2, the subspace $\mathcal{H}(b_2)$ has a basis $\{f_1, \ldots, f_\ell\}$, where $f_i = (1 - \bar{a}_i z)^{-1}$, $i = 1, \ldots, \ell$. Note that this is not an orthonormal basis.

Lemma 3 *An orthonormal basis for $\mathcal{H}(b_2)$ is given by $\{\hat{f}_1, \ldots, \hat{f}_\ell\}$, where*

$$[\hat{f}_1(z) \;\; \hat{f}_2(z) \;\; \ldots \;\; \hat{f}_\ell(z)] = [f_1(z) \;\; f_2(z) \;\; \ldots \;\; f_\ell(z)] \; \Lambda^{-1/2}$$

with $\Lambda^{-1/2}$ defined from

$$\Lambda = \left[\frac{1}{1 - \overline{a}_i a_j}\right]_{1 \le i,j \le \ell},$$

in such a way that $\Lambda^{-1/2}\Lambda\Lambda^{-1/2} = \mathrm{I}$ and $\Lambda^{-1/2} = (\Lambda^{-1/2})^*$.

Proof. First note that Λ is well defined because $a_i \neq a_k$ for $i \neq k$, and that $\Lambda = \Lambda^*$; moreover $\Lambda > 0$, since

$$\begin{aligned} 0 < \|\sum_i \xi_i f_i\|_2^2 &= \sum_{i,k} \xi_i \overline{\xi}_k \frac{1}{2\pi} \int_0^{2\pi} \frac{f_i(e^{j\theta})}{1 - a_k e^{-j\theta}} d\theta \\ &= \sum_{i,k} \xi_i \overline{\xi}_k \frac{1}{1 - a_k \overline{a}_i} = [\overline{\xi}_1 \ldots \overline{\xi}_\ell] \Lambda \begin{bmatrix} \xi_1 \\ \vdots \\ \xi_\ell \end{bmatrix}. \end{aligned}$$

So such $\Lambda^{-1/2}$ exists. Let e_i denote the vector $[0 \cdots 0 \; 1 \; 0 \cdots 0]^T \in \mathrm{R}^\ell$, where 1 is in the i th position. Then, note that

$$\begin{aligned} \langle \hat{f}_i \, , \, \hat{f}_k \rangle &= \langle [f_1 \ldots f_\ell]\Lambda^{-1/2} e_i \, , \, [f_1 \ldots f_\ell]\Lambda^{-1/2} e_k \rangle \\ &= \frac{1}{2\pi} \int_0^{2\pi} e_k^T \Lambda^{-1/2} \begin{bmatrix} \frac{f_1(e^{j\theta})}{1-a_1 e^{-j\theta}} & \cdots & \frac{f_\ell(e^{j\theta})}{1-a_1 e^{-j\theta}} \\ \vdots & & \vdots \\ \frac{f_1(e^{j\theta})}{1-a_\ell e^{-j\theta}} & \cdots & \frac{f_\ell(e^{j\theta})}{1-a_\ell e^{-j\theta}} \end{bmatrix} \Lambda^{-1/2} e_i d\theta \\ &= e_k^T \Lambda^{-1/2} \Lambda \Lambda^{-1/2} e_i \\ &= e_k^T e_i. \quad \square \end{aligned}$$

Lemma 4 *For any $h \in \mathcal{H}^2(\mathbb{D})$ we have*

$$(\mathbf{P}_{\mathcal{H}(b_2)}h)(z) = [f_1(z) \ \cdots \ f_\ell(z)] \ \Lambda^{-1} \begin{bmatrix} h(a_1) \\ \vdots \\ h(a_\ell) \end{bmatrix}. \tag{2.4}$$

Proof. The result can be seen from the following set of equalities:

$$\begin{aligned} \mathbf{P}_{\mathcal{H}(b_2)}h &= \langle h \, , \, \widehat{f}_1 \rangle \, \widehat{f}_1 + \cdots + \langle h \, , \, \widehat{f}_\ell \rangle \, \widehat{f}_\ell \\ &= [f_1 \ \cdots \ f_\ell] \ \Lambda^{-1/2} \begin{bmatrix} \langle h \, , \, \widehat{f}_1 \rangle \\ \vdots \\ \langle h \, , \, \widehat{f}_\ell \rangle \end{bmatrix} \\ &= [f_1 \ \cdots \ f_\ell] \ \Lambda^{-1/2} \Lambda^{-1/2} \begin{bmatrix} \frac{1}{2\pi} \int_0^{2\pi} \frac{h(e^{j\theta})}{1-a_1 e^{-j\theta}} d\theta \\ \vdots \\ \frac{1}{2\pi} \int_0^{2\pi} \frac{h(e^{j\theta})}{1-a_\ell e^{-j\theta}} d\theta \end{bmatrix} \\ &= [f_1 \ \cdots \ f_\ell] \ \Lambda^{-1} \begin{bmatrix} h(a_1) \\ \vdots \\ h(a_\ell) \end{bmatrix}. \quad \square \end{aligned}$$

Now we can define the *compressed shift* operator, denoted by $\mathbf{T}$, associated with $\mathcal{H}(m)$ as follows $\mathbf{T} : \mathcal{H}(m) \rightarrow \mathcal{H}(m)$, and $\mathbf{T}g = \mathbf{P}_{\mathcal{H}(m)}\mathbf{S}g$, for any $g \in \mathcal{H}(m)$. More precisely

$$\begin{aligned} (\mathbf{T}g)(z) &= (\mathbf{P}_{\mathcal{H}(m)}\mathbf{S}g)(z) \\ &= \mathbf{P}_{\mathcal{H}(m)}zg(z) \\ &= zg(z) - m(z)\phi_{-1} \end{aligned} \tag{2.5}$$

where $\phi_{-1} \in \mathbb{C}$ is as in (2.2). The adjoint of $\mathbf{T}$ is $\mathbf{T}^* = \mathbf{S}^*|_{\mathcal{H}(m)}$. Notice that $\mathbf{T}\mathbf{P}_{\mathcal{H}(m)} = \mathbf{P}_{\mathcal{H}(m)}\mathbf{S}$, i.e. $\mathbf{T}$ and $\mathbf{S}$ "intertwine" the projection $\mathbf{P}_{\mathcal{H}(m)}$.

More generally, let $f \in \mathcal{H}^\infty(\mathbb{D})$ be an arbitrary function, and $\mathbf{M}_f : \mathcal{H}^2(\mathbb{D}) \rightarrow \mathcal{H}^2(\mathbb{D})$ the multiplication operator induced by f. Then we define

$$f(\mathbf{T}) := \mathbf{P}_{\mathcal{H}(m)}\mathbf{M}_f|_{\mathcal{H}(m)}.$$

Again we have the intertwining property $f(\mathbf{T})\mathbf{P}_{\mathcal{H}(m)} = \mathbf{P}_{\mathcal{H}(m)}\mathbf{M}_f$.

Remarks.

(i) In a certain precise sense, $h(\mathbf{T})$ may be regarded as the operator gotten by formally substituting $\mathbf{T}$ for z in the power series expansion of $h(z)$:

$$h(\mathbf{T}) = \sum_{k=0}^{\infty} h_k \mathbf{T}^k = \mathbf{P}_{\mathcal{H}(m)} h(\mathbf{S})|_{\mathcal{H}(m)}. \tag{2.6}$$

This will be true for all of the functions which we will consider.

(ii) In systems theory, it is well-known that an operator $\mathbf{A} : \mathcal{H}^2(\mathsf{D}) \to \mathcal{H}^2(\mathsf{D})$ commutes with the shift $\mathbf{S} : \mathcal{H}^2(\mathsf{D}) \to \mathcal{H}^2(\mathsf{D})$ if and only if $\mathbf{A} = \mathbf{M}_f$ for some $f \in \mathcal{H}^\infty(\mathsf{D})$. This is the mathematical statement that every stable time-invariant input-output operator admits a transfer function. In this case, $\|\mathbf{A}\| = \|f\|_\infty$, where $\|\mathbf{A}\|$ denotes the operator norm of $\mathbf{A}$ (see Section 2.3.1, and [54]). We will present a generalization of this fact in Section 2.9.1.

(iii) For any $g \in \mathcal{H}(m)$ we can compute $y = h(\mathbf{T})g$ as follows:

$$y(z) = (\mathbf{P}_{\mathcal{H}(m)} hg)(z) = h(z)g(z) - m(z)(\mathbf{P}_+ m^* hg)(z)$$

$$= h(z)g(z) - m(z)[1 \;\; z \;\; z^2 \;\; \cdots] \begin{bmatrix} h_1 & h_2 & h_3 & \cdots \\ h_2 & h_3 & h_4 & \cdots \\ h_3 & h_4 & h_5 & \cdots \\ \vdots & \vdots & \vdots & \ddots \end{bmatrix} \begin{bmatrix} \phi_{-1} \\ \phi_{-2} \\ \phi_{-3} \\ \vdots \end{bmatrix}, \tag{2.7}$$

where ϕ_{-i}'s are as in (2.2). If $h(z)$ is a polynomial, then $h_k = 0$ for all $k \geq K+1$ for some $K \in \mathbf{Z}_+$, and in this case $y(z)$ can be easily computed from $h(z)$, $g(z)$, $m(z)$ and $\phi_{-1}, \ldots, \phi_{-K}$. Also note that $m(\mathbf{T}) = \mathbf{0}$, the zero operator. It is easy to see this from (2.6) which implies that $m(\mathbf{T})g = \mathbf{P}_{\mathcal{H}(m)} m(\mathbf{S})g = \mathbf{P}_{\mathcal{H}(m)} mg$ for any $g \in \mathcal{H}(m)$, but $mg \in m\mathcal{H}^2(\mathsf{D})$, so $\mathbf{P}_{\mathcal{H}(m)} mg = 0$. Thus for all $g \in \mathcal{H}(m)$ we have $m(\mathbf{T})g = 0$.

(iv) In fact, if for some $v \in \mathcal{H}^\infty$ we have $v(\mathbf{T}) = \mathbf{0}$, then $v = mg$ for some $g \in \mathcal{H}^\infty$. Indeed if $v(\mathbf{T}) = \mathbf{0}$ then $\mathbf{P}_{\mathcal{H}}v = \mathbf{P}_{\mathcal{H}}v(\mathrm{S})1 = v(\mathbf{T})\mathbf{P}_{\mathcal{H}}1$, which implies that $v \in m\mathcal{H}^2$ so $v = mg$ with $|v| = |g|$; hence $g \in \mathcal{H}^\infty$.

(v) As we have mentioned above when $m(z)$ is *rational*, $\mathcal{H}(m)$ is a *finite dimensional* subspace of $\mathcal{H}^2(\mathrm{D})$. Then, in this case since $\mathbf{T}$ is defined from $\mathcal{H}(m)$ to $\mathcal{H}(m)$, it is a *finite dimensional linear operator*, i.e. given a basis for $\mathcal{H}(m)$ we can express $\mathbf{T}$ as a square matrix of size equal to the dimension of $\mathcal{H}(m)$. For example, let us consider $m(z) = b_1(z) = z^n$, then $\{1,\ z, \ldots, z^{n-1}\}$ is an orthonormal basis for $\mathcal{H}(b_1)$. In this basis $\mathbf{T}$ (shift compressed to $\mathcal{H}(b_1)$) can be represented by an $n \times n$ matrix:

$$(\mathbf{T}g)(z) = \begin{bmatrix} 1 & z & z^2 & \ldots & z^{n-1} \end{bmatrix} \begin{bmatrix} 0 & 0 & \cdots & 0 \\ 1 & 0 & \cdots & 0 \\ 0 & \ddots & \ddots & \vdots \\ 0 & 0 & 1 & 0 \end{bmatrix} \begin{bmatrix} g_0 \\ g_1 \\ \vdots \\ g_{n-1} \end{bmatrix}$$

where $g(z) = g_0 + g_1 z + g_2 z^2 + \cdots + g_{n-1} z^{n-1} \in \mathcal{H}(b_1)$. Similarly, the compressed shift on $\mathcal{H}(b_2)$ can be represented by an $\ell \times \ell$ matrix, using the projection formula given by Lemma 4, see Section 2.9 below.

2.8 Hankel and Toeplitz operators

Hankel and Toeplitz operators play an important role in the solution of $\mathcal{H}^\infty$ control problems. These operators are defined as follows.

Given a function $v \in \mathcal{L}^\infty(\mathrm{T})$ with two sided Fourier series expansion

$$v(e^{j\theta}) = \sum_{k=-\infty}^{\infty} v_k e^{jk\theta},$$

the Hankel, $\mathbf{\Gamma}_v$, and Toeplitz, $\mathbf{\Upsilon}_v$, operators with symbol v are defined as follows; $\mathbf{\Gamma}_v : \mathcal{H}^2(\mathrm{D}) \to \mathcal{L}^2(\mathrm{T}) \ominus \mathcal{H}^2(\mathrm{D})$, $\mathbf{\Upsilon}_v : \mathcal{H}^2(\mathrm{D}) \to \mathcal{H}^2(\mathrm{D})$,

$$\mathbf{\Gamma}_v f = \mathbf{P}_- v f$$

$$\Upsilon_v f = \mathbf{P}_+ v f.$$

for any $f \in \mathcal{H}^2(\mathbf{D})$. In terms of Fourier coefficients f_k, $k = 0, 1, \ldots$, of $f(z)$ we can express Hankel and Toeplitz operators as infinite matrices:

$$(\Gamma_v f)(z) = [z^{-1}\ z^{-2}\ z^{-3}\ \ldots] \begin{bmatrix} v_{-1} & v_{-2} & v_{-3} & \cdots \\ v_{-2} & v_{-3} & v_{-4} & \cdots \\ v_{-3} & v_{-4} & v_{-5} & \cdots \\ \vdots & \vdots & \vdots & \ddots \end{bmatrix} \begin{bmatrix} f_0 \\ f_1 \\ f_2 \\ \vdots \end{bmatrix}$$

and

$$(\Upsilon_v f)(z) = [1\ \ z\ \ z^2\ \ \ldots] \begin{bmatrix} v_0 & v_{-1} & v_{-2} & \cdots \\ v_1 & v_0 & v_{-1} & \cdots \\ v_2 & v_1 & v_0 & \cdots \\ \vdots & \vdots & \vdots & \ddots \end{bmatrix} \begin{bmatrix} f_0 \\ f_1 \\ f_2 \\ \vdots \end{bmatrix}.$$

We can represent the shift and the compressed shift operators in terms of the Hankel and Toeplitz operators. For example we have already seen that the action of the operator $h(\mathbf{T})$ on an element $g \in \mathcal{H}(m)$ can be computed from (2.7), which involves an infinite size Hankel matrix. We should note that this infinite size Hankel matrix is finite rank if $h(z)$ is rational.

It is also easy to see that if $h \in \mathcal{H}^\infty(\mathbf{D})$ we have

$$(h(\mathbf{S})f)(z) = h(z)f(z) = (\Upsilon_h f)(z).$$

So, the multiplication operator with symbol $h \in \mathcal{H}^\infty(\mathbf{D})$ can be seen as the Toeplitz operator with the same symbol.

2.9 Generalized interpolation

In this section, we would like to discuss a powerful operator-theoretic approach to interpolation, which is due to the mathematician Donald Sarason [89]. Sarason's theorem will allow us to do interpolation theory for infinite dimensional SISO systems. In later chapters of this book we will apply this result to two benchmark problems consisting of delay systems and flexible beams.

We will also discuss, a far-reaching generalization due to Sz. Nagy and Foias known as the *commutant lifting theorem* [94] which will allow us to solve the general ("standard") $\mathcal{H}^\infty$ optimization problem for distributed multivariable plants.

In this section, all of our Hardy spaces $\mathcal{H}^p(\mathbf{D})$, $1 \leq p \leq \infty$, will be defined on the unit disc $\mathbf{D}$ in the standard way. We will also consider the Hardy spaces $\mathcal{H}^p(\mathbf{C}_+)$, $1 \leq p \leq \infty$, defined on the right half plane $\mathbf{C}_+$ in the standard way (see [94]).

In order to motivate Sarason's theorem, note that in order to solve the $\mathcal{H}^\infty$-optimal sensitivity problem, we are required to solve the following type of mathematical problem (see Chapter 4). Compute

$$\gamma_{opt} := \inf\{\|L - UV\|_\infty : V \in H^\infty(\mathbf{D})\} \tag{2.8}$$

where $L, U \in \mathcal{H}^\infty(\mathbf{D})$, and U is inner.

The precise problem we would like to address here is finding a way of computing $\mathcal{H}^\infty$ optimal performance γ_{opt}, and also finding the corresponding optimal V_{opt}. This will also give us an operator theoretic derivation of the Nevanlinna-Pick interpolation theorem.

2.9.1 Sarason's Theorem

Let us now consider the problem defined by (2.8). Given L and U we want to find γ_{opt} and V_{opt}. First we can define the shift invariant subspace $U\mathcal{H}^2(\mathbf{D})$ and its orthogonal complement $\mathcal{H}(U)$. Let $\mathbf{T}$ denote the compressed shift operator associated with $\mathcal{H}(U)$. Now we are ready to state Sarason's theorem.

Theorem 5 (D. Sarason) *Let* $\mathbf{A} : \mathcal{H}(U) \to \mathcal{H}(U)$ *be any (bounded linear) operator such that* $\mathbf{TA} = \mathbf{AT}$. *Then there exists a function* $f \in \mathcal{H}^\infty(\mathbf{D})$, *such that* $\mathbf{A} = f(\mathbf{T})$, *and* $\|A\| = \|f\|_\infty$.

The proof of Sarason's theorem is far beyond the scope of this Chapter. Let us just say here that there are numerous proofs now of this result, and even of the more general **commutant lifting theorem.** For a very readable discussion of these results see Chapter 8 and [28].

What we will do here now is show how Sarason's theorem leads to a solution for the computation of

$$\gamma_{opt} = \inf\{\|L - UV\|_\infty : V \in \mathcal{H}^\infty(\mathbf{D})\}.$$

Consider the operator $L(\mathbf{T}) = \mathbf{P}_{\mathcal{H}(U)}\mathbf{M}_L|_{\mathcal{H}(U)}$. Clearly, $L(\mathbf{T})$ commutes with $\mathbf{T}$. Therefore from Sarason's theorem there exists a function $L_{opt} \in \mathcal{H}^\infty(\mathbf{D})$ such that $L_{opt}(\mathbf{T}) = L(\mathbf{T})$ and $\|L(\mathbf{T})\| = \|L_{opt}\|_\infty$. Now since

$$\mathbf{P}_{\mathcal{H}(U)}\mathbf{M}_L|_{\mathcal{H}(U)} = \mathbf{P}_{\mathcal{H}(U)}\mathbf{M}_{L_{opt}}|_{\mathcal{H}(U)},$$

we have that

$$L_{opt} = L - UV_{opt}$$

for some $V_{opt} \in \mathcal{H}^\infty(\mathbf{D})$. We now claim that for every $V \in \mathcal{H}^\infty(\mathbf{D})$

$$\|L - UV\|_\infty \geq \|L(\mathbf{T})\|. \tag{2.9}$$

Since $\|L - UV_{opt}\|_\infty = \|L(\mathbf{T})\|$, we must have (assuming the claim) that

$$\gamma_{opt} = \inf_{V \in \mathcal{H}^\infty(\mathbf{D})} \|L - UV\|_\infty = \|L(\mathbf{T})\|_\infty = \|L - UV_{opt}\|_\infty.$$

In order to prove (2.9), note that we can write (in a unique manner)

$$L - UV = L_1 + L_2$$

where $L_1 \in \mathcal{H}(U)$, and $L_2 \in U\mathcal{H}^2(\mathbf{D})$. Therefore

$$\|L - UV\|_\infty \geq \|L_1\|_\infty. \tag{2.10}$$

Now $L_1 = \mathbf{P}_{\mathcal{H}(U)}L$, and therefore

$$\|L_1\|_\infty = \|\mathbf{P}_{\mathcal{H}(U)}\mathbf{M}_L\| \geq \|\mathbf{P}_{\mathcal{H}(U)}\mathbf{M}_L|_{\mathcal{H}(U)}\| = \|L(\mathbf{T})\|. \tag{2.11}$$

Combining (2.10) and (2.11), we get (2.9) as claimed.

We will discuss a procedure for computing $\|L(\mathbf{T})\|$ for certain infinite dimensional systems in Chapter 4. Notice however that we can reduce the $\mathcal{H}^\infty$-optimization problem to the computation of the norm of an explicit operator $L(\mathbf{T})$ which we will call the *Sarason operator.* In Chapter 4 we will see that the Sarason operator is equivalent to a Hankel operator.

We should note that all of the preceding results derived in the unit disc $\mathbf{D}$ are valid as well in the right-half plane $\mathbf{C}_+$. Namely, we have that for $L, U \in \mathcal{H}^\infty(\mathbf{C}_+)$ with U inner,

$$\inf\{\|L - UV\|_\infty : V \in \mathcal{H}^\infty(\mathbf{C}_+)\} = \|\mathbf{P}_{\mathcal{H}(U)}\mathbf{M}_L|_{\mathcal{H}^2(\mathbf{C}_+)\ominus U\mathcal{H}^2(\mathbf{C}_+)}\|$$

where $\mathbf{P}_{\mathcal{H}(U)} : \mathcal{H}^2(\mathbf{C}_+) \to \mathcal{H}^2(\mathbf{C}_+) \ominus U\mathcal{H}^2(\mathbf{C}_+)$ is orthogonal projection.

Returning to the unit disc, we would like to discuss a procedure for the computation of V_{opt}. We will *assume* that

$$\|L(\mathbf{T})\|^2 = \sigma(L(\mathbf{T})^*L(\mathbf{T}))$$

where $\sigma(L(\mathbf{T})^*L(\mathbf{T}))$ is the largest eigenvalue of $L(\mathbf{T})^*L(\mathbf{T})$. This assumption is valid for $L(\mathbf{T})$ finite dimensional, and will hold for all the Sarason operators whose norm is strictly greater than its essential norm.

Let $v \in \mathcal{H}(U)$ be a *maximal vector* for $L(\mathbf{T})$, i.e. v is a non-zero eigenvector of $L(\mathbf{T})^*L(\mathbf{T})$ corresponding to the largest singular value $\sigma(L(\mathbf{T})^*L(\mathbf{T}))$. Then we have

Theorem 6 *With the above notation,*

$$L_{opt} = L - UV_{opt} = (L(\mathbf{T})v)/v.$$

Moreover, L_{opt} has the form $\|L(\mathbf{T})\|$ times an inner function.

Proof. We have by definition $\|L_{opt}\|_\infty = \|L(\mathbf{T})\|$, and $L_{opt}(\mathbf{T}) = L(\mathbf{T})$. In what follows below, $\| \ \|_2$ denotes the 2-norm on $\mathcal{H}^2(\mathsf{D})$, and $\| \ \|$ the operator norm as before. Note that

$$\|L(\mathbf{T})v\|_2^2 = \langle L(\mathbf{T})v, L(\mathbf{T})v\rangle = \langle L(\mathbf{T})^* L(\mathbf{T})v, v\rangle = \|L(\mathbf{T})\|^2 \|v\|_2^2.$$

Hence $\|L(\mathbf{T})v\|_2 = \|L(\mathbf{T})\| \, \|v\|_2$. Now

$$\begin{aligned} \|L(\mathbf{T})\| \, \|v\|_2 &= \|L(\mathbf{T})v\|_2 = \|\mathbf{P}_{\mathcal{H}(U)} L_{opt} v\|_2 \\ &\leq \|L_{opt} v\|_2 \leq \|L_{opt}\|_\infty \, \|v\|_2 = \|L(\mathbf{T})\| \, \|v\|_2. \end{aligned}$$

Thus $\mathbf{P}_{\mathcal{H}(U)} L_{opt} v = L_{opt} v$, and so the modulus of L_{opt} cannot be less than $\|L(\mathbf{T})\|$ on a set of positive measure. Hence $L_{opt} = L(\mathbf{T})v/v$, and is of the form $\|L(\mathbf{T})\|$ times an inner function, completing the proof. □

From the proof of Theorem 6 we also have that V_{opt} is unique. This is not true in general for the $\mathcal{H}^\infty$-optimization problem in the multiple input/multiple output case [28].

2.9.2 Nevanlinna-Pick Theorem

In this section, we would like to indicate how Sarason's theorem implies the Nevanlinna-Pick theorem. This will give a completely operator theoretic proof to a result that was originally derived from complex analysis [41]. The method is extremely important since it can be used to give interpolation results in the matrix (and operator) case which allows us to extend the Nevanlinna-Pick framework to multiple input/multiple output distributed systems. The section however is optional, and what follows is independent of the results here. In what follows , unless stated to the contrary all the Hardy spaces will be defined on the unit disc $\mathbf{D}$. We set $\mathcal{H}^p := \mathcal{H}^p(\mathbf{D})$, $1 \le p \le \infty$.

The problem of Nevanlinna-Pick concerns finding necessary and sufficient conditions for the existence of an analytic $\psi : \mathbf{D} \to \mathbf{D}$, such that $\psi(a_i) = b_i$, $i = 1, \ldots, n$. We assume that the a_i are distinct. We now put the Nevanlinna-Pick problem into the Sarason framework [89].

Accordingly, set

$$B(z) = \prod_{i=1}^{n} \left(\frac{z - a_i}{1 - \overline{a_i} z} \right).$$

Let $\mathcal{H} := \mathcal{H}^2 \ominus B\mathcal{H}^2$. Note that $B(z)$ is in the form $b_2(z)$ of Section 2.6. Hence and by Lemmas 2 and 3

$$g_i(z) := \frac{1}{1 - \overline{a_i} z} \quad \text{for } i = 1, \ldots, n.$$

form a basis for $\mathcal{H}$.

We need to understand how the compressed shift $\mathbf{T} : \mathcal{H} \to \mathcal{H}$ acts on the basis $\{g_1, \ldots, g_n\}$. Actually, it is a bit simpler to consider $\mathbf{T}^* : \mathcal{H} \to \mathcal{H}$, the adjoint of $\mathbf{T}$.

Lemma 5 $\quad \mathbf{T}^* g_i = \overline{a_i} g_i$

for $i = 1, \ldots, n$, *i.e. the* g_i*'s are eigenvectors of* $\mathbf{T}^*$.

Proof. Let ζ be a point on the unit circle, i.e., suppose that $|\zeta| = 1$. It is easy to check that

$$\mathbf{T}^* g(\zeta) = \overline{\zeta}(g(\zeta) - g(0))$$

for $g \in \mathcal{H}$. But

$$\overline{\zeta}(g_i(\zeta) - g_i(0)) = \overline{\zeta} \left(\frac{1}{1 - \overline{a_i}\zeta} - 1 \right) = \overline{\zeta} \left(\frac{\overline{a_i}\zeta}{1 - \overline{a_i}\zeta} \right) = \overline{a_i} g_i$$

since $\zeta\overline{\zeta} = 1$. □

Note that an operator $\mathbf{A} : \mathcal{H} \to \mathcal{H}$ commutes with $\mathbf{T}$ if and only if $\mathbf{A}^*$ commutes with $\mathbf{T}^*$. But from Sarason's theorem $\mathbf{A}$ commutes with $\mathbf{T}$ if and only if there exists a function $\psi \in \mathcal{H}^\infty(\mathbf{D})$ such that $\mathbf{A} = \psi(\mathbf{T})$. If ψ is such that $\psi(\mathbf{T}) = \mathbf{A}$, then we say that ψ *interpolates* $\mathbf{A}$. The reason for this terminology should become clear from the next lemma.

Lemma 6 *Let $\mathbf{A} : \mathcal{H} \to \mathcal{H}$ be defined by $\mathbf{A}^* g_i = \overline{b_i} g_i$ for $i = 1, \ldots, n$. (Clearly, $\mathbf{A}^*$ commutes with $\mathbf{T}^*$, and so by the above $\mathbf{A}$ commutes with $\mathbf{T}$.) Then $\psi \in \mathcal{H}^\infty(\mathbf{D})$ interpolates $\mathbf{A}$ (i.e. $\mathbf{A} = \psi(\mathbf{T})$) if and only if $\psi(a_i) = b_i$ for $i = 1, \ldots, n$.*

Proof. Suppose $\mathbf{A} = \psi(\mathbf{T})$. Then by definition $\mathbf{A} g_i = \psi g_i - B q_i$ for some $q_i \in \mathcal{H}^2$, $i = 1, \ldots, n$. Hence

$$(\mathbf{A} g_i)(a_i) = \psi(a_i) g_i(a_i). \tag{2.12}$$

On the other hand,

$$(\mathbf{A}^* g_i)(a_i) = \overline{b_i} g_i(a_i).$$

But

$$(\mathbf{A} g_i)(a_i) = \langle \mathbf{A} g_i, g_i \rangle = \langle g_i, \mathbf{A}^* g_i \rangle = \langle g_i, \overline{b_i} g_i \rangle = b_i g_i(a_i),$$

and so by (2.12), $\psi(a_i) = b_i$ as required.

Conversely suppose that $\psi(a_i) = b_i$, $i = 1, \ldots, n$. Then we claim that $\psi(\mathbf{T}) = \mathbf{A}$. Indeed, we need only check this on the basis vectors g_i. Hence, we must show that $\psi(\mathbf{T})^* g_i = \overline{b_i} g_i$, $i = 1, \ldots, n$. Since the g_i form a basis, it is enough to show that

$$\langle \psi(\mathbf{T})^* g_i, g_j \rangle = \langle \overline{b_i} g_i, g_j \rangle \quad \forall i, j. \tag{2.13}$$

In order to do this, first note that

$$\psi(\mathbf{T})g_i = \psi g_i - Bq_i$$

for some $q_i \in \mathcal{H}^2(\mathsf{D})$, $i = 1, \dots, n$. Moreover,

$$\langle g_i, Bq_j \rangle = 0$$

for $1 \leq i, j \leq n$ since $g_i \perp B\mathcal{H}^2(\mathsf{D})$. Consequently,

$$\begin{aligned}
\langle \psi(\mathbf{T})^* g_i, g_j \rangle &= \langle g_i, \psi(\mathbf{T}) g_j \rangle \\
&= \langle g_i, \psi g_j - Bq_j \rangle \\
&= \langle g_i, \psi g_j \rangle \\
&= \overline{\langle \psi g_j, g_i \rangle} \\
&= \overline{\psi(a_i) g_j(a_i)} \\
&= \overline{\psi(a_i)} g_i(a_j) \quad \text{since } g_i(a_j) = \overline{g_j(a_i)}.
\end{aligned}$$

Now $\langle \overline{b_i} g_i, g_j \rangle = \overline{b_i} g_i(a_j)$. Since $\psi(a_i) = b_i$, we have (2.13). □

We are now ready to give the alternate proof of the Nevanlinna-Pick theorem:

Theorem 7 (Nevanlinna-Pick) *Notation as above. Then there exists an analytic $\psi : \mathsf{D} \to \overline{\mathsf{D}}$ such that $\psi(a_i) = b_i$ for $i = 1, \dots, n$ if and only if the associated Pick matrix Q is positive semi-definite, i.e.*

$$Q := \left[\frac{1 - \overline{b_i} b_j}{1 - \overline{a_i} a_j} \right]_{i,j=1,\dots,n} \geq 0.$$

Proof. By Lemma 6 and Sarason's theorem, there exists an interpolating function ψ such that $\|\psi\|_\infty \leq 1$ (i.e., $\psi : \mathsf{D} \to \overline{\mathsf{D}}$) if and only if $\|\mathbf{A}\| \leq 1$, where $\mathbf{A} = \psi(\mathbf{T})$. Therefore we must write out necessary and sufficient conditions for $\mathbf{A} : \mathcal{H} \to \mathcal{H}$, $\mathbf{A}^* g_i = \overline{b_i} g_i$ to have norm less than 1. But $\|\mathbf{A}\| \leq 1 \iff \|\mathbf{A}^*\| \leq 1 \iff$

$$\langle \mathbf{A}^* g, \mathbf{A}^* g \rangle \leq \langle g, g \rangle \quad \forall g \in \mathcal{H}. \tag{2.14}$$

Now any $g \in \mathcal{H}$ has the unique representation

$$g = \alpha_1 g_1 + \cdots + \alpha_n g_n. \tag{2.15}$$

Then

$$\begin{aligned} \langle g, g \rangle &= \sum_{1 \leq i,j \leq n} \alpha_i \overline{\alpha_j} \langle g_i, g_j \rangle \\ &= \sum_{1 \leq i,j \leq n} \alpha_i \overline{\alpha_j} g_i(a_j) \\ &= \sum_{1 \leq i,j \leq n} \alpha_i \overline{\alpha_j} \frac{1}{1 - \overline{a_i} a_j}. \end{aligned}$$

But

$$\mathbf{A}^* g = \alpha_1 \overline{b_1} g_1 + \cdots + \alpha_n \overline{b_n} g_n \tag{2.16}$$

and so

$$\langle \mathbf{A}^* g, \mathbf{A}^* g \rangle = \sum_{1 \leq i,j \leq n} \alpha_i \overline{\alpha_j}\, \overline{b_i}\, b_j \left(\frac{1}{1 - \overline{a_i} a_j} \right). \tag{2.17}$$

Thus we have (2.14) if and only if

$$\sum_{1 \leq i,j \leq n} \alpha_i \overline{\alpha_j}\, \overline{b_i}\, b_j \left(\frac{1}{1 - \overline{a_i} a_j} \right) \leq \sum_{1 \leq i,j \leq n} \alpha_i\, \overline{\alpha_j} \left(\frac{1}{1 - \overline{a_i} a_j} \right).$$

In other words, (2.14) is equivalent to the condition

$$0 \leq \sum_{1 \leq i,j \leq n} \alpha_i \overline{\alpha_j} \left(\frac{1 - \overline{b_i} b_j}{1 - \overline{a_i} a_j} \right) .$$

Since the α_i are arbitrary, this last condition is equivalent to the positive-definiteness of the Pick matrix Q, from which we can conclude the proof of the theorem. □

Remark. Note that $\ker Q \neq \{0\}$ if and only if there exists $g \in \mathcal{H}$ such that $\|\mathbf{A}^* g\| = \|g\| = 1$ (which in turn equivalent to $\|\mathbf{A}\| = \|\mathbf{A}^*\|$). Let $f = (\mathbf{A}\mathbf{A}^* g)/\|\mathbf{A}\mathbf{A}^* g\|$. Then $\|\mathbf{A}f\| = \|f\| = 1$, and thus from

$$\begin{aligned} 1 = \|f\| = \|\mathbf{A}f\| &= \|\psi(\mathbf{T})f\| \\ &= \|\mathbf{P}_{\mathcal{H}}(\psi f)\| \leq \|\psi f\| \leq \|f\| \end{aligned}$$

we infer $\psi f = \mathbf{A}f$ and

$$\frac{1}{2\pi} \int (1 - \|\psi\|^2) |f|^2 d\theta = \|f\|^2 = \|\psi f\|^2 = 1 .$$

whence since $f \neq 0$, $1 \geq |\psi|^2$ a.e. on $\mathbf{T}$ we can conclude with

$$\psi = \mathbf{A}f/f \quad , \quad |\psi|^2 = 1 \text{ a.e. on } \mathbf{T}.$$

That is, ψ is and inner function, and ψ is unique.

2.10 Riemann Mapping Theorem

In the next chapter we will define gain and phase margin problems, and solve them later using Nevanlinna-Pick interpolation theory. Before discussing the control problems of interest, we will need to state (without proof) a classical result from complex function theory that will be the key in transforming one optimal solution to another in the various problems being considered. See also [96].

It is well-known in control that the right half plane $\mathbb{C}_+$ and the unit disc $\mathbf{D}$ are analytically or conformally equivalent. This means that there exists a 1-1 analytic map $\phi : \mathbb{C}_+ \to \mathsf{D}$ with analytic inverse. Actually in this case it is easy to show that ϕ must have the form

$$\phi(s) = \exp(j\theta)\,\frac{s-\bar{b}}{s+b}, \quad b \in \mathbb{C}_+\,,\ \theta \in [0\,,\,2\pi]\,.$$

An important problem in complex analysis consists in identifying regions of $\mathbb{C}$ which are conformally equivalent to D. This will also be crucial for us in solving the stability margin optimization problem. In order to answer this question we will need the following definition:

Definition. Let $\mathsf{G} \subset \mathbb{C}$ be a region. Then G is said to be *simply connected* if it has no holes.

The key result that we need is:

Theorem 8 (Riemann Mapping Theorem) *Let $\mathsf{G} \subset \mathbb{C}$ be any simply connected region (not equal to $\mathbb{C}$ itself). Then G is conformally equivalent to the unit disc D.*

We should note here that all known proofs of the Riemann Mapping Theorem are non-constructive. However there are some excellent approximation procedures available. An important (and fortunate) fact is that for the regions that appear in the classical control problems, one can easily find the required conformal equivalences.

Examples.

(a) Let

$$\mathsf{G} := \mathbb{C}\backslash\{s \in \mathbb{C} : s \text{ real } \ s \le 0\}.$$

It is easy to see that G is simply connected. Now we want to construct a conformal equivalence $\phi : \mathsf{G} \to \mathsf{D}$. This we do in two stages. First we construct $\phi_1 : \mathsf{G} \to \mathbb{C}_+$. This is given by

$\phi_1(s) := \sqrt{s}$. Next we construct $\phi_2 : \mathbb{C}_+ \to \mathbf{D}$. This is given by $\phi_2(s) := (s-1)/(s+1)$. Then we set

$$\phi(s) := \phi_2 \circ \phi_1(s) = \frac{\sqrt{s}-1}{\sqrt{s}+1}.$$

(b) This example will be important for us in connection with the gain margin problem. Let

$$\mathsf{G} := \mathbb{C}\backslash\{(-\infty, \frac{a}{a-1}] \cup [\frac{b}{b-1}, \infty)\}, \quad b > 1 > a > 0.$$

Then we claim that

$$\phi(s) := \frac{1 - [(1-(\frac{b-1}{b})s)/(1-(\frac{a-1}{a})s]^{1/2}}{1 + [(1-(\frac{b-1}{b})s)/(1-(\frac{a-1}{a})s]^{1/2}}$$

is a conformal equivalence $\phi : \mathsf{G} \to \mathbf{D}$ such that $\phi(0) = 0$. It is a general fact, for given simply connected region G with $0 \in \mathsf{G}$, a conformal equivalence of G with $\mathbf{D}$ which maps 0 to 0 is unique up to rotation of $\mathbf{D}$, that is if $\phi^{(i)} : \mathsf{G} \to \mathbf{D}$ $i = 1, 2$, are two conformal equivalences with $\phi^{(i)}(0) = 0$, then

$$\exp(j\theta)\phi^{(1)} = \phi^{(2)}$$

for some $\theta \in [0, 2\pi)$. If we then require that ϕ have real coefficients (such a conformal equivalence will always exist in all the cases of interest to us), then ϕ is uniquely determined.

For the case at hand, we now show that ϕ is the required conformal equivalence. As above we construct it by stages.

(1) $\phi_1 : \mathsf{G} \to \mathbb{C}\backslash(-\infty, 0]$. Indeed let $\phi_1(s) := (\alpha - s)/(s - \beta)$, with $\alpha = \frac{a}{a-1}$ and $\beta = \frac{b}{b-1}$. Then $\phi_1(\alpha) = 0$, $\phi_1(\beta) = \infty$ and $\phi_1(0) = \frac{-\alpha}{\beta}$.

(2) $\phi_2 : \mathbb{C}\backslash(-\infty, 0] \to \mathbb{C}_+$, $\phi_2(s) := \sqrt{s}$. Note $\phi_2(\frac{-\alpha}{\beta}) = \sqrt{\frac{-\alpha}{\beta}}$.

(3) $\phi_3 : \mathbb{C}_+ \to \mathsf{D}$,

$$\phi_3(s) := \frac{s - \sqrt{\frac{-\alpha}{\beta}}}{s + \sqrt{\frac{-\alpha}{\beta}}}.$$

Then $\phi = \phi_3 \circ \phi_2 \circ \phi_1$ is the required map.

(c) We now want to construct the conformal equivalence $\phi : \mathsf{G} \to \mathsf{D}$ with $\phi(0) = 0$ where

$$\mathsf{G} := \mathbb{C} \backslash \{\frac{s}{s-1} : s = \exp j\theta,\ \theta \in [-\theta_1, \theta_1],\ \theta_1 \in (0, \pi]\}.$$

We will need this in our solution of the phase margin problem. We will just summarize the mappings. Full details may be found in [16]. Set

$$\begin{aligned} a &:= \frac{\sin\theta_1}{1-\cos\theta_1}, \\ c &:= \sqrt{\frac{1-j/a}{1+j/a}}. \end{aligned}$$

Then one may show that the required conformal equivalence is given by

$$\phi = \phi_4 \circ \phi_3 \circ \phi_2 \circ \phi_1,$$

where

$$\begin{aligned} \phi_1(s) &:= s - 1/2, \\ \phi_2(s) &:= js, \\ \phi_3(s) &:= \sqrt{\frac{1+2s/a}{1-2s/a}}, \\ \phi_4(s) &:= \frac{s-c}{s+\bar{c}}. \end{aligned}$$

2.11 Remarks on Nevanlinna-Pick Interpolation

In this section, we will discuss some aspects of Nevanlinna-Pick interpolation theory which are most relevant to control.

The classical Nevanlinna-Pick problem is concerned with finding necessary and sufficient conditions for the existence of an analytic function $f : \mathbf{D} \to \overline{\mathbf{D}}$ such that $f(a_i) = \tilde{b}_i$, $1 \leq i \leq n$, for given points $a_i \in \mathbf{D}$, $\tilde{b}_i \in \overline{\mathbf{D}}$. The famous Nevanlinna-Pick theorem then asserts that such an interpolating f exists if and only if the following *Pick* matrix

$$Q := \left[\frac{1 - \tilde{b}_i\overline{\tilde{b}_j}}{1 - a_i\overline{a_j}}\right]_{1 \leq i,j \leq n}$$

is positive semi-definite.

The problems involved in stability margin optimization depend on a slight variation of this set-up. Let $a_1, \ldots, a_n \in \mathbf{D}$, $b_1, \ldots, b_n \in \mathbf{C}$. Let $\gamma > 0$. Then we want to find necessary and sufficient conditions for the existence of an analytic $f_\gamma : \mathbf{D} \to \mathbf{D}$ such that $f_\gamma(a_i) = \gamma^{-1} b_i$, $i = 1, \ldots, n$. Using Nevanlinna-Pick interpolation theory, it is easy to find the minimal γ, γ_{opt}, for which this problem is solvable. Indeed, set

$$A := \left[\frac{1}{1 - a_i\overline{a_j}}\right]_{1 \leq i,j \leq n},$$

$$B := \left[\frac{b_i\overline{b_j}}{1 - a_i\overline{a_j}}\right]_{1 \leq i,j \leq n}.$$

Then f_γ exists if and only if

$$A - (1/\gamma^2)B \geq 0.$$

The proof of Theorem 7 implies that $A > 0$. Using this, and the fact that $A^{-1}B$ and $A^{-\frac{1}{2}}BA^{-\frac{1}{2}}$ have the same eigenvalues, it is easy to show that

$$\gamma_{opt} = \sqrt{\lambda_{max}},$$

where λ_{max} is the largest eigenvalue of $A^{-1}B$.

The interpolation data associated to such an interpolation problem (with f_γ), we will write in matricial form as

$$\begin{matrix} a_1 & a_2 & \dots & a_n \\ b_1 & b_2 & \dots & b_n \end{matrix}.$$

2.11.1 Suboptimal interpolants

In order to compute the optimal and suboptimal compensators relative to various optimization problems, we will have to give a formula for f_γ. We will only sketch this here referring the reader to [62], [60] for all the details. Using a Moebius transformation if necessary, we may assume without lose of generality that the a_i's are non-zero. Let $\gamma_{opt} < \gamma$. Then the Pick matrix Q_γ associated to the interpolation data a_i, $\gamma^{-1}b_i$ $(1 \leq i \leq n)$ is positive definite. In order to write down the interpolating functions (that is all the functions satisfying all the interpolation conditions), we will need a bit of notation.

Set

$$B(z) := \prod_{i=1}^{n} \frac{a_i - z}{1 - \overline{a_i}z} \cdot \frac{a_i}{|a_i|}$$

$$y_i := \overline{B(0)}/\overline{a_i},\; y^T := [y_1\; y_2\; \dots\; y_n]$$

where y^T is the transpose of y. Let $x := {Q_\gamma}^{-1}y$. Next set

$$\begin{aligned} P(z) &:= \overline{B(0)}B(z) - \sum_{i=1}^{n} \frac{B(z)}{z - a_i}x_i \\ Q(z) &:= -z\left(\sum_{i=1}^{n} \frac{\gamma^{-1}\overline{b_i}}{1 - \overline{a_i}z}\overline{x_i}\right) \\ \tilde{P}(z) &:= B(z)\overline{P(1/\overline{z})} \\ \tilde{Q}(z) &:= B(z)\overline{Q(1/\overline{z})}. \end{aligned}$$

Then all solutions to our interpolation problem are given by

$$f_\gamma(z) = \frac{\tilde{P}(z)g(z) + \tilde{Q}(z)}{P(z) + Q(z)g(z)}$$

where $g : \mathsf{D} \to \overline{\mathsf{D}}$ is an arbitrary analytic function. Notice that the parametrization of the interpolants has the form of a *linear fractional transformation* with free parameter g, and with $P, \tilde{P}, Q, \tilde{Q}$ completely determined by the interpolation data.

Now without loss of generality choose $\gamma = 1$ and consider the difference

$$\begin{aligned} \Delta &:= |P(z) + Q(z)g(z)|^2 - |\tilde{P}(z)g(z) + \tilde{Q}(z)|^2 \\ &= |P(z)|^2 + |Q(z)|^2|g(z)|^2 + 2\mathrm{Re}P(z)\overline{Q(z)g(z)} \\ &- |\tilde{Q}(z)|^2 - |\tilde{P}(z)|^2|g(z)|^2 - 2\mathrm{Re}\tilde{Q}(z)\overline{\tilde{P}(z)g(z)} \end{aligned} \tag{2.18}$$

for $|z| = 1$. By definition we have

$$P(z)\overline{Q(z)} - \tilde{Q}(z)\overline{\tilde{P}(z)} = P(z)\overline{Q(z)} - \overline{Q(1/\bar{z})\overline{P(1/\bar{z})}} = 0$$

Then it is easy to see that

$$\Delta = (|P(z)|^2 - |\tilde{Q}(z)|^2) - (|\tilde{P}(z)|^2 - |Q(z)|^2)|g(z)|^2 \tag{2.19}$$

$$= (|P(z)|^2 - |Q(z)|^2)(1 - |g(z)|^2) \tag{2.20}$$

Again, using the definitions of $P(z)$ and $Q(z)$ it can be shown that

$$|P(z)|^2 - |Q(z)|^2 = |B(0)|^2 + y^TQ^{-T}y \geq |B(0)|^2.$$

Thus,

$$\begin{aligned} \Delta &= |P(z) + Q(z)g(z)|^2 - |\tilde{P}(z)g(z) + \tilde{Q}(z)|^2 \\ &= (|B(0)|^2 + y^TQ^{-T}y)(1 - |g(z)|^2) \end{aligned} \tag{2.21}$$

The above fact will be used later in the extension of the Nevanlinna-Pick theorem to the case where interpolation data can be on T as well as inside D.

2.11.2 Optimal interpolant

We have solved the problem of finding f_γ in case $\gamma > \gamma_{opt}$. Let us consider now the *degenerate case* in which $\gamma = \gamma_{opt}$. It is this case which corresponds to the optimal solution in various $\mathcal{H}^\infty$–optimization problems. If $\gamma = \gamma_{opt}$, then the Pick matrix Q_γ is singular, and there is a unique function $f_\gamma : \mathsf{D} \to \mathsf{D}$ such that $f_\gamma(a_i) = \gamma^{-1}b_i$. This will correspond to the optimal solution in the various optimization problems we will be considering. It turns out that this function is an *all-pass*, i.e. a Blaschke product times a constant [41]. We will now use the above parametrization of the suboptimal solutions to find this unique all-pass function.

Indeed, let m be the rank of Q_γ, $m < n$. After suitable re-ordering of the a_i, without loss of generality we may assume that the top left $m \times m$ principal minor M_γ of Q_γ is nonsingular. Consider the restricted interpolation problem of finding all analytic functions $h_\gamma : \mathsf{D} \to \mathsf{D}$ such that $h_\gamma(a_i) = \gamma^{-1}b_i$ for $i = 1, \ldots m$. Then M_γ is the Pick matrix associated to this restricted problem, which is nonsingular. Thus we can find as above $P_o(z), Q_o(z), \tilde{P}_o(z), \tilde{Q}_o(z)$ for this restricted problem such that all solutions h_γ are given by

$$h(z) = \frac{\tilde{P}_o(z)g(z) + \tilde{Q}_o(z)}{P_o(z) + Q_o(z)g(z)}$$

where $g : \mathsf{D} \to \overline{\mathsf{D}}$ is an arbitrary analytic function. Now we must choose g such that h_γ satisfies the rest of the interpolation conditions, i.e

$$h_\gamma(a_j) = \gamma_{opt}^{-1} b_j$$

for $j = m+1, \ldots, n$. Hence g must satisfy

$$\gamma_{opt}^{-1} b_j = \frac{\tilde{P}_o(a_j)g(a_j) + \tilde{Q}_o(a_j)}{P_o(a_j) + Q_o(a_j)g(a_j)} \tag{2.22}$$

for $j = m+1, \ldots, n$.

Since rank $Q_\gamma < n$, it is a standard fact from Nevanlinna-Pick theory that there is a unique constant g_o with $|g_o| = 1$, such that

$g(z) \equiv g_o$ is the only function which satisfies the above requirements [62]. Consequently, the unique solution to the degenerate problem for $\gamma = \gamma_{opt}$ is

$$f_{\gamma_{opt}} = \frac{\tilde{P}_o(z)g_o + \tilde{Q}_o(z)}{P_o(z) + Q_o(z)g_o}.$$

Now if a Nevanlinna-Pick problem is solvable, then one can always find a solution which is an all-pass [41]. Since $f_{\gamma_{opt}}$ is unique, $f_{\gamma_{opt}}$ is an all-pass.

2.11.3 An Extension of γ_{opt}

In most cases in practical control problems, one has to consider boundary interpolations as well. We will see below for example that when P is strictly proper, in order to guarantee the properness and hence the causality of a compensator derived from interpolation, we are forced into this situation. (See also [60] and [96].) Thus the problem we will be interested in is the following. Let $a_1, \ldots, a_{n-r} \in \mathbf{D}$, $a_{n-r+1}, \ldots, a_n \in \mathbf{T}$ ($\mathbf{T}$:= the unit circle), and $b_1, \ldots, b_n \in \mathbf{C}$. Then once again we want to find the minimal γ, $\hat{\gamma}_{opt}$, for which there exists an analytic $f_\gamma : \mathbf{D} \to \mathbf{D}$ with $f_\gamma(a_i) = \gamma^{-1} b_i$, $i = 1, \ldots, n$. Fortunately $\hat{\gamma}_{opt}$ is very easy to compute. We will represent the interpolation data relative to this problem as

$$\begin{array}{cccccc} a_1 & \cdots & a_{n-r} & a_{n-r+1} & \cdots & a_n \\ b_1 & \cdots & b_{n-r} & b_{n-r+1} & \cdots & b_n \end{array}$$

Now define

$$\hat{\gamma}_{opt} := \min\left(\gamma_{opt}, \frac{1}{b_{n-r+1}}, \frac{1}{b_{n-r+2}}, \ldots, \frac{1}{b_n}\right)$$

where γ_{opt} is defined relative to the interior interpolation data $a_1, \ldots, a_{n-r}$, $b_1, \ldots, b_{n-r}$. Then we have the following elementary result:

Theorem 9 *There exists an analytic function $f_\gamma : \overline{\mathsf{D}} \to \mathsf{D}$ such that $f_\gamma(a_i) = \gamma^{-1} b_i$ for $i = 1, \ldots, n$ if and only if $\gamma > \hat{\gamma}_{opt}$.*

Proof. Since $\gamma > \gamma_{opt}$, we have that there exists $h_\gamma : \mathsf{D} \to \mathsf{D}$ such that $h_\gamma(a_j) = \gamma^{-1} b_j$ for $j = 1, \ldots, n - r$. But from our discussion in Section 2.11, there exist rational functions, analytic in D, completely determined by the first $n - r$ pairs of the interpolation data, $P, \tilde{P}, Q, \tilde{Q}$, such that

$$h_\gamma = \frac{\tilde{P} g + \tilde{Q}}{P + Q g} \tag{2.23}$$

where $g : \mathsf{D} \to \overline{\mathsf{D}}$ is an arbitrary analytic function. We need therefore to find g so that the additional boundary interpolation conditions $h_\gamma(a_{n-r+i}) = \gamma^{-1} b_{n-r+i}$ for $i = 1, \ldots, r$ are satisfied. But from equation 2.23 we have

$$g = \frac{\tilde{Q} - P h_\gamma}{-\tilde{P} + Q h_\gamma}.$$

Thus the additional interpolation conditions are satisfied if and only if

$$g(a_{n-r+i}) = \frac{\tilde{Q}(a_{n-r+i}) - P(a_{n-r+i})\gamma^{-1} b_{n-r+i}}{-\tilde{P}(a_{n-r+i}) + Q(a_{n-r+i})\gamma^{-1} b_{n-r+i}} =: \eta_i$$

for $i = 1, \ldots, r$. Now (2.21) implies that

$$0 < \frac{(|B(0)|^2 + y^T Q^{-T} y)(1 - |\eta_i|^2)}{|P(a_{n-r+i}) + Q(a_{n-r+i}) g(a_{n-r+i})|^2} \Longrightarrow |\eta_i|^2 < 1 \,.$$

Hence such a g always exists. Indeed note that $a_{n-r+i} \in \mathsf{T}$, $\eta_i \in \mathsf{D}$, $i = 1, \ldots, r$. Then for $\epsilon > 0$, set $\mathsf{D}_\epsilon := \{|z| < 1 + \epsilon\}$. Computing the corresponding Pick matrix for functions $g : \mathsf{D}_\epsilon \to \mathsf{D}$ satisfying the latter interpolation conditions, it is easy to check that for ϵ sufficiently small, the matrix will be positive definite, and hence the interpolation problem will always have a solution. □

We will see below that $\hat{\gamma}_{opt}$ can be identified with a fundamental control-theoretic invariant which will allow us to solve the stability margin optimization problems, and even the robust stabilization and weighted sensitivity minimization problems.

Chapter 3

Issues in Feedback Control

3.1 Closed loop stability

Consider the feedback control system shown in Figure 3.1, where P denotes the plant to be controlled, and C represents the controller to be designed. The external signals are r, d, v, n, reference, actuator disturbance, output disturbance, and measurement noise, respectively. The internal signals of interest are e, u, y, measured error, command input, and the plant output, respectively.

For the benefit of some readers let us recall the meaning of the feedback connections shown in Figure 3.1:

$$y = v + Pu \ , \ e = r - (n + y) \ , \ u = d + Ce \tag{3.1}$$

Then, in order for the system to be well defined we must have

$$y = v + P(d + C(r - n - y)) \implies (I + PC)y = v + Pd + PC(r - n) \ .$$

In other words, the inverse of $(I + PC)$ must be well defined.

We will assume that the energy contents of the external signals are finite, i.e. these signals belong to $\mathcal{L}^2(\mathsf{R}_+)$. We will consider linear time-invariant systems P and C represented by their transfer functions $P(s)$

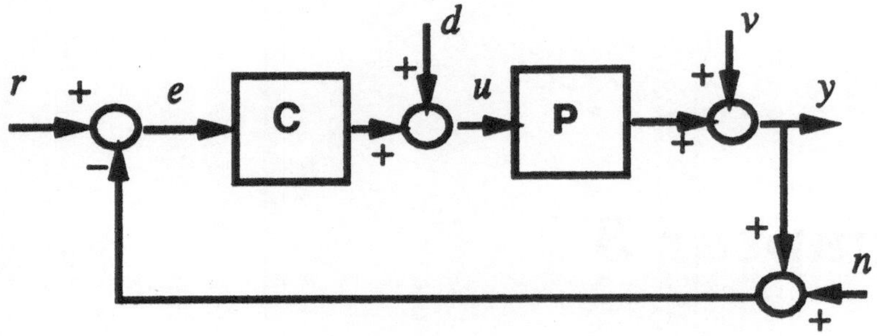

Figure 3.1: Closed Loop System

and $C(s)$, respectively. (The precise assumptions on these transfer functions will be given below.) Throughout the book, we consider plants and controllers whose transfer functions can be represented as ratios of two $\mathcal{H}^\infty$ functions, i.e. $P(s) = P_n(s)/P_d(s)$ and $C(s) = C_n(s)/C_d(s)$ for some $P_n, P_d, C_n, C_d \in \mathcal{H}^\infty$.

Definition: Let $C(s) = 0$ and $v = 0$ in Figure 3.1, i.e., the system is open loop and the output is $y = Pu = Pd$. Then, we say that the *open loop system P is stable* if it is a bounded linear operator from $\mathcal{L}^2(\mathsf{R}_+)$ to $\mathcal{L}^2(\mathsf{R}_+)$; that is for every $u \in \mathcal{L}^2(\mathsf{R}_+)$ the output $y = Pu$ is in $\mathcal{L}^2(\mathsf{R}_+)$ and

$$\sup_{u\neq 0} \frac{\|y\|_2}{\|u\|_2} < \infty. \tag{3.2}$$

One can interpret this definition as follows: The plant P is stable if all finite energy command signals u give rise to finite energy outputs y, and the maximum energy amplification from u to y is finite.

In view of Sections 2.2 and 2.4 the stability of P is equivalent to having its transfer function $P(\cdot)$ in $\mathcal{H}^\infty$. In fact, the maximum energy amplification, given by (3.2), is the norm of this operator, which is equal to the $\mathcal{H}^\infty$ norm of the transfer function, i.e., $P(s) = y(s)/u(s)$ and

$$\sup_{u\neq 0} \frac{\|y\|_2}{\|u\|_2} = \operatorname{ess\,sup}_{\omega\in\mathsf{R}} |P(j\omega)| = \|P\|_\infty.$$

So, from a control system theoretic point of view $\mathcal{H}^\infty$ can be seen as the set of transfer functions of all stable systems.

The above definition of stability is an input/output stability concept in the sense of bounded energy amplification. There are several other stability definitions in the case of distributed parameter systems, see e.g. [9], [111], [115]. In the finite dimensional systems case all these definitions lead to the "usual" definition of stability.

Definition: The *closed loop system* $[C, P]$ *is stable* if all transfer functions (from any external input to any internal signal) are in $\mathcal{H}^\infty$.

This definition means that the closed loop system stability is equivalent to the following: all finite energy external inputs give rise to finite energy internal signals and the maximum energy amplification in the system is finite. It is easy to see from Figure 3.1 and the algebraic relations (3.1) that all transfer functions can be expressed in terms of the following four functions:

$$\begin{aligned}
S(s) &:= (1 + P(s)C(s))^{-1}, \\
T(s) &:= 1 - S(s) = P(s)C(s)(1 + P(s)C(s))^{-1}, \\
C(s)S(s) &= C(s)(1 + P(s)C(s))^{-1}, \\
P(s)S(s) &= P(s)(1 + P(s)C(s))^{-1}.
\end{aligned}$$

For example, the sensitivity function S is the transfer function from v to y, or from r to e; the complementary sensitivity function T is the transfer function from n to y, or from r to y; PS is the transfer function from d to y; CS is the transfer function from r to u; etc. Therefore, the closed loop system is stable if and only if all four transfer functions S, T, CS, PS are in $\mathcal{H}^\infty$

3.2 Controller parametrization

In a controller design the most important requirement is stability of the closed loop system $[C, P]$, with controller C and plant P. If $[C, P]$ is a stable closed loop system, then we say that the *controller C stabilizes* *the plant P*. In this section we will characterize the set of all stabilizing controllers for a given plant.

We will assume that P has a factorization of the form $P(s) = N(s)/D(s)$, where $N, D \in \mathcal{H}^\infty$ such that there exist $X, Y \in \mathcal{H}^\infty$ satisfying the Bezout identity

$$N(s)X(s) + D(s)Y(s) = 1\,. \tag{3.3}$$

If such a factorization holds for P then N, D are called *coprime factors* of P. In fact existence of such a factorization is necessary for the existence of a stabilizing controller; see [91].

Theorem 10 *([91] [116]): A controller C, which is a ratio of two $\mathcal{H}^\infty$ functions, stabilizes the plant P if and only if C is in the form*

$$C(s) = \frac{X(s) + D(s)Q(s)}{Y(s) - N(s)Q(s)}, \tag{3.4}$$

where $Q \in \mathcal{H}^\infty$ is the free parameter to be chosen according to design specifications, other than stabilization. □

Unless otherwise stated, the following assumption on the plant will be in effect.

Assumption 3.1 (On the plant): We consider the following class of SISO, LTI, possibly infinite dimensional plants:

$$P(s) = \frac{M_n(s)N_1(s)N_2(s)}{M_d(s)}, \quad s \in \mathbb{C}_+,$$

which when transformed to $z-$plane (via a conformal map $z = \frac{s-a}{s+a}$, with $a > 0$) $P(s)$ become

$$p(z) = \frac{m_n(z)n_1(z)n_2(z)}{m_d(z)}, \quad z \in \mathbf{D},$$

where m_n is an arbitrary (possibly infinite dimensional) inner function, m_d is a rational inner function, n_1 is possibly an infinite dimensional outer function with $n_1^{-1} \in \mathcal{H}^\infty(\mathbf{D})$, and n_2 is a rational outer function. We will assume that $P(j\omega)$ is continuous on $j\mathbf{R}_e$ except at finitely many points. We will also assume that m_n has finitely many essential singularities, and that $m_n(0) \neq 0 \neq m_d(0)$. □

The key restrictions in this assumption are that n_2 and m_d are rational. Rationality of n_2 is a technical requirement, which can be relaxed if we allow improper controllers (see [67] for an example). Having m_d rational means that the plant can have only finitely many unstable modes. This assumption is crucial in our solution of the $\mathcal{H}^\infty$ control problems, since the complexity of computations depend on the order of $m_d(z)$. Another restriction is that the plant does not have any poles on the imaginary axis (i.e., m_d is inner). The theory presented here does not need this assumption, but we will use it to avoid technical details involving "outer factor absorption" problem for which we refer to [24]. The assumption that $m_n(0) \neq 0 \neq m_d(0)$ is without loss of generality, because we can always choose the parameter a in the conformal map in such a way that this is satisfied (i.e. a is chosen such that $P(s)$ does not have any pole or zero at $s = a$).

Example: An infinite dimensional plant example satisfying Assumption 3.1 is

$$P(s) = \frac{e^{-hs}(s-0.05)}{(s+1)(s+0.1-e^{-h_1 s})}, \qquad h_1 = 2\ln(\frac{5}{3}), \qquad h > 0.$$

Note that the only point in $\overline{\mathbb{C}_+}$ where the term $(s+0.1-e^{-h_1 s})$ becomes zero is $s = 0.5$. So, the plant has only one pole in the closed right half plane. We can easily check that the multiplicity of this pole is 1. Therefore, in this example we can identify the components of P as follows

$$\begin{aligned} M_n(s) &= e^{-hs}\frac{s-0.05}{s+0.05} \\ M_d(s) &= \frac{s-0.5}{s+0.5} \end{aligned}$$

$$
\begin{aligned}
N_1(s) &= \frac{(s-0.5)(s+0.05)}{(s+0.1-e^{-h_1 s})(s+0.5)} \\
N_2(s) &= \frac{1}{s+1}.
\end{aligned}
$$

Note that $N_1^{-1} \in \mathcal{H}^\infty$, in fact

$$
N_1(0.5)^{-1} = (\frac{1}{0.55})(1 + \frac{6}{5}\ln(\frac{5}{3})), \text{ and } N_1(\infty)^{-1} = 1.
$$

One can use a conformal map in order to transform these functions to the z−domain.

Note that in the above example the choice of N_1 and N_2 is not unique. For example we could have chosen

$$
\begin{aligned}
\tilde{N}_1(s) &= \frac{(s-0.5)(s+0.05)}{(s+0.1-e^{-h_1 s})(s+1)} \\
\tilde{N}_2(s) &= \frac{1}{s+0.5},
\end{aligned}
$$

instead of the above N_1 and N_2, and still satisfy Assumption 3.1.

For an arbitrary plant whose transfer function is a ratio of two $\mathcal{H}^\infty$ functions, e.g. $P = N/D$ with $N, D \in \mathcal{H}^\infty(\mathbb{C}_+)$, we can check if it satisfies Assumption 3.1 as follows. First of all we need to look at the zeros of $D(s)$ in the right half plane, there should be finitely many, and none on the imaginary axis. Then, we can obtain the Bode magnitude plot for $20\log|P(j\omega)|$, if it converges to a finite number as $\omega \to \infty$, (i.e. if P is not strictly proper) then the assumption is satisfied, we can choose $N_2 = 1$. If this plot converges to $-\infty$ as $\omega \to \infty$, then the rate of decay has to be an integer multiple of −20 dB per decade, i.e. as $\omega \to \infty$ the function has to "look like" the Bode magnitude plot of a rational function. This is necessary to have a rational $N_2 \in \mathcal{H}^\infty(\mathbb{C}_+)$ and to have $N_1^{-1} \in \mathcal{H}^\infty(\mathbb{C}_+)$.

We can obtain a characterization of stabilizing controllers for the plant given in the above example by using Theorem 10 and solving the corresponding Bezout equation as shown below.

Example: Let us consider the plant in the example just given. We can define $N(s) = M_n(s)N_1(s)N_2(s)$ and $D(s) = M_d(s)$. Then, the Bezout equation can be solved as follows. Note that

$$Y(s) = \frac{1 - N(s)X(s)}{D(s)}.$$

Since $D(s)$ has a single zero in the closed right half plane (at $s = 0.5$), we need to find an $X \in \mathcal{H}^\infty(\mathbb{C}_+)$ such that $X(0.5) = N(0.5)^{-1}$. Therefore, we can simply choose

$$X(s) = N(0.5)^{-1} = e^{h/2}(\frac{10}{3} + 4\ln(\frac{5}{3})).$$

In general $X(s)$ is constructed from the interpolation conditions $X(p_i) = N(p_i)^{-1}$ where $p_i \in \mathbb{C}_+$ are the zeros of $D(s)$ (i.e. poles of $P(s)$) with multiplicity 1. If a pole p_i has multiplicity $k \geq 2$ then we also require $(\frac{\partial^i}{\partial s^i})N(s)X(s)|_{s=p_i} = 0$ for all $i = 1, \ldots, k-1$. There are finitely many interpolation conditions, if the plant has finitely many right half plane poles. In such cases $X(s)$ can always be chosen as a rational function; in fact by Lagrange interpolation $x(z)$ can be chosen as a polynomial of degree $(\ell - 1)$, where ℓ is the dimension of M_d. On the other hand note that, when X and D are rational and N is infinite dimensional, Y is infinite dimensional.

3.3 Robust stability

In the above discussion, we have assumed that the plant transfer function is given by $P(s)$, and we have characterized the set of all controllers stabilizing this *fixed plant*. However, usually $P(s)$ is a *nominal* representation of the *actual* plant, whose transfer function is denoted by $P_\Delta(s)$. The part of P_Δ which does not appear in P is called the *unmodeled dynamics*. There are a number of ways to represent the unmodeled dynamics, e.g. multiplicative, additive, coprime factor perturbations of the nominal plant.

Multiplicative Perturbation:

$$P_\Delta(s) = P_m(s) = P(s)\ (1 + \Delta_m(s)),$$

Additive Perturbation:

$$P_\Delta(s) = P_a(s) = P(s) + \Delta_a(s),$$

Coprime Factor Perturbation:

$$P_\Delta(s) = P_{cf}(s) = \frac{N(s) + \Delta_N(s)}{D(s) + \Delta_D(s)}\ , \quad \Delta_{cf}(s) = [\Delta_N(s)\ \ \Delta_D(s)],$$

where $P(s) = N(s)/D(s)$ is the nominal plant and $\Delta(s)$ is the unmodeled dynamics. In general $\Delta(s)$ is unknown, but a ***frequency dependent upper bound function*** $W(s)$ (called the ***uncertainty weight***) can be introduced to represent the uncertainty in the form

$$\|\Delta(j\omega)\| < |W(j\omega)| \quad \forall\ \omega \in \mathbf{R} \tag{3.5}$$

for example

$$1 \ < \ \operatorname*{ess\ sup}_{\omega} \left| \frac{W_m(j\omega)}{\Delta_M(j\omega)} \right| , \tag{3.6}$$

$$1 \ < \ \operatorname*{ess\ sup}_{\omega} \left| \frac{W_a(j\omega)}{\Delta_a(j\omega)} \right| , \tag{3.7}$$

$$1 \ < \ \operatorname*{ess\ sup}_{\omega} \frac{|W_{cf}(j\omega)|^2}{|\Delta_N(j\omega)|^2 + |\Delta_D(j\omega)|^2}\ , \tag{3.8}$$

where $W_m(s), W_a(s), W_{cf}(s)$ are known functions. The actual plant is in the form P_Δ, with Δ unknown; but the nominal plant P and the uncertainty weight W, satisfying (3.5), are known. We will assume that in the case of multiplicative and additive perturbations the actual plant P_Δ and the nominal plant P have the same number of right half plane poles; when dealing with coprime factor perturbations we will relax this condition.

If $P_a = P_m$ then W_m and W_a have to satisfy:

$$|W_m(j\omega)P(j\omega)| > |\Delta_a(j\omega)| \tag{3.9}$$

and

$$\left|\frac{W_a(j\omega)}{P(j\omega)}\right| > |\Delta_m(j\omega)|. \tag{3.10}$$

Our assumptions on the nominal plant imply that

$$|P(j\omega)| = |N_1(j\omega)N_2(j\omega)|.$$

This means that we can see $W_m N_1 N_2$ as an additive uncertainty bound, and $\frac{W_a}{N_1 N_2}$ as a multiplicative uncertainty bound. Therefore, if $W_a = W_m N_1 N_2$, then whether we consider multiplicative or additive perturbations does not make any difference.

Assumption 3.2 (On the weights): We assume that W_{cf}, W_{cf}^{-1}, $(W_m N_2)$, $(W_m N_2)^{-1}$, are rational functions in $\mathcal{H}^\infty$, and the additive uncertainty weight is given by

$$W_a = W_m N_1 N_2. \tag{3.11}$$

We further assume that there exist ω_o and $K \geq 2$, such that

$$|W_m(j\omega)| > K \quad \text{for all} \quad \omega \geq \omega_o. \tag{3.12}$$

This assumption implies that W_m must be improper, whenever N_2 is strictly proper, in which case (3.12) is automatically satisfied. The purpose of choosing W_a in the form (3.11) is to make the additive uncertainty problem the same as the multiplicative uncertainty problem.

Considering the unmodeled dynamics in the plant we require that the controller C, which is fixed, stabilizes not only the nominal plant P but also all possible plants P_Δ with Δ satisfying (3.5), for a given weight W. If a controller meets this requirement then we say that C *robustly stabilizes the plant.* Necessary and sufficient conditions for a controller C to robustly stabilize the plant are given by the following.

Theorem 11 *Consider the classes of plants described by P_m, P_a and P_{cf}, with the nominal plant $P = N/D$ (where N, D are coprime, and P_m, P_a and P have the same number of right half plane poles), and the uncertainty bounds given by (3.6), (3.7) and (3.8), for some weights W_a, W_m and W_{cf}, respectively. Suppose that the nominal plant satisfies Assumption 3.1, and the weights satisfy Assumption 3.2. Let C be a controller, which is a ratio of two $\mathcal{H}^\infty$ functions, such that $C(j\omega)$ is continuous on $j\mathbb{R}_e$ except at finitely points, and C stabilizes the nominal plant P. When dealing with P_a or P_m we also assume that C has finitely many poles in the closed right half plane. Then, C robustly stabilizes the plant if and only if*

case (i): multiplicative perturbations

$$\|W_m PC(1+PC)^{-1}\|_\infty \leq 1, \tag{3.13}$$

case (ii): additive perturbations

$$\|W_a C(1+PC)^{-1}\|_\infty \leq 1, \tag{3.14}$$

case (iii): coprime factor perturbations

$$\left\| W_{cf} D^{-1} \begin{bmatrix} (1+PC)^{-1} \\ C(1+PC)^{-1} \end{bmatrix} \right\|_\infty \leq 1. \tag{3.15}$$

Proof. For the coprime factor perturbations case, see [42]. (This was first shown by [110] for the finite dimensional case.) Since W_a satisfies (3.11) P_m is the same as P_a, and

$$\|W_m PC(1+PC)^{-1}\|_\infty = \|W_a C(1+PC)^{-1}\|_\infty.$$

Therefore, the multiplicative perturbation case is the same as the additive perturbation case, and it is sufficient to prove the theorem for either of these cases. For the sufficiency part, the proof of [10] goes through because P and C have finitely many poles in the closed right half plane, and (3.12) implies that as $\omega \to \infty$ the Nyquist plot of $P(j\omega)C(j\omega)$ lies within the unit circle. This can be seen from the following:

$$PC = \frac{PC(1+PC)^{-1}}{1-PC(1+PC)^{-1}}.$$

Then, $|PC| < 1$ if

$$|PC(1+PC)^{-1}| < \frac{1}{2}$$

This means that $|PC| < 1$ if

$$|W_m PC(1+PC)^{-1}| < \frac{|W_m|}{2}.$$

On the other hand, by Assumptions 3.1 and 3.2, $|W_m(j\omega)| > 2$ and $P(j\omega)C(j\omega)$ is continuous for all $\omega \geq \omega_o'$, (for some $\omega_o' < \infty$). Hence, we have that

$$\|W_m PC(1+PC)^{-1}\|_\infty \leq 1$$

implies $|P(j\omega)C(j\omega)| < 1$ for all $\omega \geq \omega_o'$. Obviously, if N_2 is strictly proper, then so is P, and hence the Nyquist plot converges to the origin as $\omega \to \infty$. But this is not necessary, a weaker condition of the form (3.12) is sufficient to apply the Nyquist criterion.

For the necessity part one has to show that if

$$\|W_m PC(1+PC)^{-1}\|_\infty > 1,$$

then there exists an admissible multiplicative perturbation Δ_m destabilizing the closed loop system. Such a perturbation can be constructed exactly the same way as in the finite dimensional case (see e.g. [16] and [18]) because of the continuity assumptions on P and C. However, this is too laborious to include in the book. □

Special Case: If P and Δ_a (or Δ_m) are stable, then in the above theorem we don't need the assumption that C has finitely many closed right half plane poles. In this case the proof goes as follows: necessity part is the same because if (3.14) is not satisfied then one can find a destabilizing perturbation. For sufficiency we note that

$$(1+P_aC)^{-1} = (1+PC)^{-1}\frac{1}{1+\Delta_a C(1+PC)^{-1}}. \tag{3.16}$$

Since C stabilizes P, $(1+PC)^{-1}$ and $C(1+PC)^{-1}$ are in $\mathcal{H}^\infty$. Therefore, when Δ_a is in $\mathcal{H}^\infty$, (3.14) and (3.7) imply that $(1+\Delta_a C(1+PC)^{-1})^{-1}$ is in $\mathcal{H}^\infty$, and hence $(1+P_aC)^{-1} \in \mathcal{H}^\infty$ for all admissible Δ_a. The identity (3.16) further implies that $C(1+P_aC)^{-1}$ and $P_a(1+P_aC)^{-1}$ are also in $\mathcal{H}^\infty$. Hence the closed loop system is stable for all admissible P_a.

Example: Consider the plant given in the previous examples, and let W_m be given as

$$W_m(s) = (s+0.2).$$

Note that $W_m(s)N_2(s) = \frac{s+0.2}{s+1}$ and $|W_m(j\omega)| \to \infty$ as $\omega \to \infty$. So, W_m satisfies the conditions of Assumption 3.2. Let the controller C be given as in Theorem 10, where N, D, X, Y are as defined before. Then, condition (3.13) becomes

$$\|W_m M_n N_1 N_2 (X + M_d Q)\|_\infty \leq 1. \tag{3.17}$$

Therefore, a *robustly stabilizing controller exists* if and only if there exists a $Q \in \mathcal{H}^\infty$ satisfying (3.17). Since M_n is inner and $W_m N_1 N_2$ is invertible in $\mathcal{H}^\infty$ by virtue of Assumptions 3.1 and 3.2, (3.17) can be reduced to

$$\|R + M_d Q_1\|_\infty \leq 1 \tag{3.18}$$

where $Q_1 = (W_m N_2) N_1 Q$ and $R = W_m N_1 N_2 X \in \mathcal{H}^\infty$. Hence a robustly stabilizing controller exists if and only if there exists a $Q_1 \in \mathcal{H}^\infty$ satisfying (3.18). Since $M_d(0.5) = 0$ the left hand side of (3.18) is greater or equal to $|R(0.5)|$ for all Q_1. On the other hand for $Q_1 = Q_{1,opt}$, defined by

$$Q_{1,opt} = \frac{R(s) - R(0.5)}{M_d(s)} \in \mathcal{H}^\infty,$$

the left hand side of (3.18) is equal to $|R(0.5)|$. Thus a robustly stabilizing controller exists if and only if $|R(0.5)| \leq 1$. Note that

$$|R(0.5)| = |W_m(0.5)N_1(0.5)N_2(0.5)X(0.5)| = \frac{0.7e^{h/2}0.55}{0.45}.$$

We conclude that a robustly stabilizing controller exists if and only if

$$h \leq 2\ln(\frac{9}{7.7}) \approx 0.312.$$

In other words if the delay is "too large," we cannot find a robustly stabilizing controller for this system, and the largest allowable delay is $2\ln(\frac{9}{7.7})$.

3.4 Robust performance

The above discussion summarized by Theorem 11 gives conditions for a controller to robustly stabilize the plant. Besides stability we are interested in the performance of the closed loop system. A typical performance condition is the sensitivity reduction, which can be stated as follows. Given a desired upper bound $W_d(s)$ for the sensitivity function (we will assume that W_d is rational and $W_d, W_d^{-1} \in \mathcal{H}^\infty$) we want to find a robustly stabilizing controller C such that

$$|(1 + P_\Delta(j\omega)C(j\omega))^{-1}| \leq |W_d(j\omega)| \quad \text{a.e.} \quad \omega \in \mathbb{R}, \tag{3.19}$$

where P_Δ represents the actual plant, which can be any transfer function of the form P_m, P_a, P_{cf}, with the nominal plant P and the uncertainty weights W_m, W_a, W_{cf}, respectively. The function $(1 + P_\Delta C)^{-1}$ is the sensitivity of the closed loop system with the actual plant P_Δ and the controller C. The problem (3.19) is called the *robust performance* problem. It is difficult to design a controller satisfying the necessary and sufficient conditions for the robust performance problem. On the other hand for the additive and multiplicative perturbation cases there is a simple sufficient condition, given by Theorem 12 below, which leads to a two block $\mathcal{H}^\infty$ controller design.

Theorem 12 *Assume that the classes of plants described by the multiplicative or the additive perturbations are as defined in Theorem 11. Consider a controller satisfying the assumptions stated in Theorem 11. Then, C solves the robust performance problem if (resp. only if) it stabilizes the nominal plant P, and satisfies the robust performance inequality*

case (i): multiplicative perturbations

$$\left\| \begin{bmatrix} W_d^{-1}(1+PC)^{-1} \\ W_m PC(1+PC)^{-1} \end{bmatrix} \right\|_\infty \leq \frac{1}{\sqrt{2}} \quad (resp. \sqrt{2}) \tag{3.20}$$

case (ii): additive perturbations

$$\left\| \begin{bmatrix} W_d^{-1}(1+PC)^{-1} \\ W_a C(1+PC)^{-1} \end{bmatrix} \right\|_\infty \leq \frac{1}{\sqrt{2}} \quad (resp. \sqrt{2}). \tag{3.21}$$

Proof. This is a consequence of Theorem 11, [16], the details are given below. A necessary condition is robust stability, which is automatically satisfied if (3.20) or (3.21) holds. Also note that by Assumption 3.2 on the weights, the problems (3.20) and (3.21) are identical. Therefore, we may consider either of these two cases. We want a stabilizing C satisfying

$$|W_d^{-1}(j\omega)(1+(P(j\omega)+\Delta_a(j\omega))C(j\omega))^{-1}| \leq 1 \tag{3.22}$$

almost everywhere on the imaginary axis, and for all Δ_a satisfying our additive uncertainty assumptions. Note that (3.22) is equivalent to having, for all admissible Δ_a,

$$|W_d^{-1}(1+PC)^{-1}| \leq |1+\Delta_a C(1+PC)^{-1}| \quad \text{a.e.} \tag{3.23}$$

(we have dropped the dependence on $(j\omega)$ for notational convenience). The inequality (3.23) is satisfied for all admissible Δ_a if and only if

$$|W_d^{-1}(1+PC)^{-1}| \leq 1-|W_a C(1+PC)^{-1}|, \quad \text{a.e.} \tag{3.24}$$

It is easy to see that condition (3.24) is satisfied if the following holds

$$|W_d^{-1}(1+PC)^{-1}|^2+|W_a C(1+PC)^{-1}|^2 \leq \frac{1}{2} \quad \text{a.e.} \tag{3.25}$$

This concludes the proof. □

The above theorem gives only a *sufficient* condition for the robust performance problem. The conservatism is in the step where we go from (3.24) to (3.25).

Remark: By Assumption 3.1 , D is equal to M_d, which is inner, so (3.15) can be reduced to

$$\left\| W_{cf} \begin{bmatrix} (1+PC)^{-1} \\ C(1+PC)^{-1} \end{bmatrix} \right\|_\infty \leq 1, \tag{3.26}$$

which looks similar to (3.21): if one chooses the weights $W_d^{-1} = W_a = W_{cf}$ then the two problems are the same. On the other hand, such a choice is possible, without violating Assumption 3.2 or rationality of W_d, only if N_1 is rational. This restricts the class of distributed plants which can be handled in problem (3.26).

3.5 Disturbance attenuation

Disturbance attenuation is another issue where we are faced with a control problem similar to (3.21) and (3.20): Consider the closed loop system shown in Figure 3.1. Let us assume that $d = r = 0$, and

$$v \in \mathcal{D}_v := \{W_v v_1 \ : \ v_1 \in \mathcal{L}^2(\mathrm{R}_+), \ \|v_1\|_2 \leq 1\}$$

$$n \in \mathcal{D}_n := \{W_n n_1 \ : \ n_1 \in \mathcal{L}^2(\mathrm{R}_+), \ \|n_1\|_2 \leq 1\},$$

where W_v and W_n are LTI "weights" shaping the frequency and magnitude of the output disturbance, and the measurement noise, respectively. A disturbance attenuation problem is to find a controller C stabilizing the closed loop system and minimizing

$$\gamma_1(C) = \sup_{v \in \mathcal{D}_v,\ n \in \mathcal{D}_n} \|y\|_2.$$

Here $y = (1 + PC)^{-1} W_v v_1 + PC(1 + PC)^{-1} W_n n_1$, therefore

$$\begin{aligned} \gamma_1(C) &= \|[(1+PC)^{-1}W_v \quad PC(1+PC)^{-1}W_n]\|_\infty \\ &= \operatorname{ess\ sup}_\omega \|[\frac{W_v(j\omega)}{1+P(j\omega)C(j\omega)} \quad \frac{W_n(j\omega)P(j\omega)C(j\omega)}{1+P(j\omega)C(j\omega)}]\| \end{aligned}$$

So, the problem is equivalent to finding a stabilizing controller which minimizes

$$\gamma_1(C) = \left\| \begin{bmatrix} W_v(1+PC)^{-1} \\ W_n PC(1+PC)^{-1} \end{bmatrix} \right\|_\infty . \tag{3.27}$$

Similarly we can define another disturbance attenuation problem by assuming $d = r = n = 0$: find a controller C, stabilizing the closed loop system and minimizing

$$\gamma_2(C) = \sup_{v \in \mathcal{D}_v} \left\| \begin{bmatrix} y \\ W_u u \end{bmatrix} \right\|_2 ,$$

where W_u is a LTI weight. It can be shown that

$$\gamma_2(C) = \left\| \begin{bmatrix} W_v(1+PC)^{-1} \\ W_u W_v C(1+PC)^{-1} \end{bmatrix} \right\|_\infty . \tag{3.28}$$

Note that the problem (3.27) is similar to the problem (3.20) and the problem (3.28) is similar to the problem (3.21), provided the weights are chosen appropriately. The major difference between the problems associated with disturbance attenuation and the problems arising in robust performance is that in Theorem 12 (for robust performance) we considered controllers with *finitely many* closed right half plane poles. However, there is no such restriction in disturbance attenuation problems. Therefore, we may first solve the disturbance attenuation problem. If the resulting controller has finitely many closed right half plane poles, then this controller also solves the robust performance problem. Otherwise, we can approximate the controller by a *finite dimensional* one, in such a way that the approximate controller stabilizes the closed loop system with infinite dimensional plant, and satisfies the robust performance inequality (3.20) (or (3.21)). This issue will be discussed in Chapter 6.

3.6 Standard $\mathcal{H}^\infty$ control problems

3.6.1 Two block problem

We conclude that several cases of robust stabilization, robust performance and disturbance attenuation problems can be solved by finding the $\mathcal{H}^\infty$ optimal controller C_{opt} from the following two-block $\mathcal{H}^\infty$ problem:

$$\gamma_{opt} := \inf_{[C,P]\ stable} \left\| \begin{bmatrix} W_1(1+PC)^{-1} \\ W_2PC(1+PC)^{-1} \end{bmatrix} \right\|_\infty , \tag{3.29}$$

where C_{opt} stabilizes the nominal plant P and achieves the $\mathcal{H}^\infty$ optimal performance γ_{opt}, i.e.

$$\gamma_{opt} := \left\| \begin{bmatrix} W_1(1+PC_{opt})^{-1} \\ W_2PC_{opt}(1+PC_{opt})^{-1} \end{bmatrix} \right\|_\infty ,$$

and W_1 and W_2 are appropriate weights related to the control problems defined above. We will assume that W_1 is real rational (i.e $W_1(s)^* =$

$W_1(s^*)$) with $W_1, W_1^{-1} \in \mathcal{H}^\infty$, and $W_2 = W_m$ is a real rational function which satisfies Assumption 3.2. Chapters 4–7 of this book deals with solutions of the two block problem defined above for SISO possibly unstable distributed plants. Extensions to multivariable and four block cases will be discussed in Chapter 8. The four block problem for MIMO plants is described below.

3.6.2 Four block problem

Note that the disturbance attenuation problems and the other $\mathcal{H}^\infty$ control problems defined above can be stated as minimizing the $\mathcal{H}^2$ norm of a certain "signal of interest," when the $\mathcal{H}^2$ norm of the external disturbance signal is normalized to unity. The signals of interest, and the external disturbances, are shaped (their magnitude an frequency distribution) by certain weighting functions. The controller generates a command signal, from the measured output signal. The weights and the plant to be controlled can be cast into a "generalized plant" G as shown in Figure 3.2, where negative feedback controller is denoted by K, ($K = -C$ for the closed loop system of Figure 3.1). In Figure 3.2 v denotes the external disturbance, z denotes the "signal of interest," and as before u is the command signal and y is the measured output.

The standard four block $\mathcal{H}^\infty$ control problem can be defined as

$$\gamma_{opt} = \inf_K \{\|z\|_2 \quad : \quad \|v\|_2 \leq 1\} \tag{3.30}$$

where the infimum is taken over all controllers stabilizing the closed loop system. It is easy to see from Theorem 1 that

$$\gamma_{opt} = \inf_K \|T_{zv}\|_\infty \tag{3.31}$$

where T_{zv} denotes the closed loop transfer function from v to z. Note that this transfer function depends on K, and whenever the plant is multi input multi output (MIMO), the weights, the controller and T_{zv} are also MIMO. The problem (3.30) is rather difficult to solve for arbitrary MIMO infinite dimensional systems. In Chapter 8 of this book

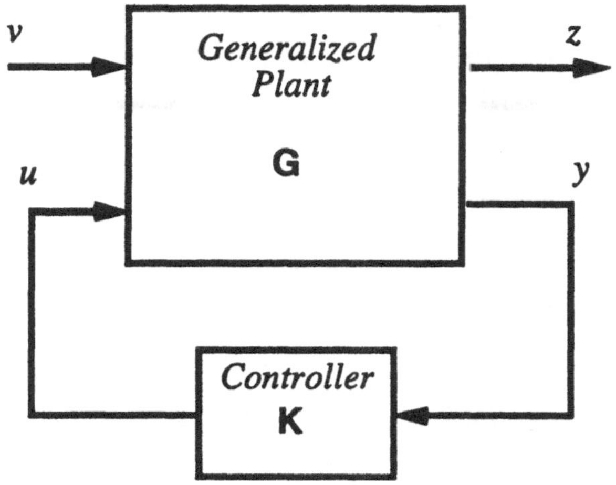

Figure 3.2: Standard Problem

we will consider this problem for stable plants and finite dimensional weights. We will show that identifying the finite and infinite dimensional parts of the problem data G simplifies the solution of this problem considerably.

3.7 Stability Margin Optimization

In this section, we will give a precise formulation to the problem of stability margin optimization and reduce it to a Nevanlinna-Pick type of interpolation problem.

Consider the following family of SISO, LTI finite dimensional plants:

$$P_k(s) = kP(s)$$

where $P(s)$ is the nominal plant transfer function, and k is a parameter taking values in $\mathsf{K} \subset \mathbb{C}$ simply connected and compact, not containing the origin, but with $1 \in \mathsf{K}$. Then we are interested in the problem of finding (if possible) a proper compensator $C(s)$ which stabilizes the closed loop system for all $k \in \mathsf{K}$. We will call this the *generalized stability margin optimization problem.*

This set-up includes all of the standard classical stability margins. Specifically, we have the following examples of the parameter variation set K ([96], [60], [16]):

(a) $\mathsf{K} = [a, b]$, $b > 1 > a > 0$. Then the above problem amounts to the classical *gain margin problem*. More precisely, if a stabilizing compensator $C(s)$ exists relative to K, then it means that $C(s)$ guarantees a gain margin of at least $20 \log b/a$ dB for the nominal plant $P(s)$.

(b) $\mathsf{K} = \{k = \exp j\theta, \theta \in [-\theta_1, \theta_1], \quad 0 < \theta_1 \leq \pi\}$. This is the *phase margin problem*. If a stabilizing compensator $C(s)$ exists relative to K, then $C(s)$ guarantees a phase margin of at least θ_1 radians for the nominal plant $P(s)$.

We should also add that there are several other possibilities for the set K corresponding to gain-phase margin, and complex parameter variations that we will not treat here. See [60], [96].

We shall now reduce these problems to one of interpolation. We now define the *sensitivity function*

$$S(s) := (1 + P(s)C(s))^{-1}.$$

Then it is very easy to show [96] that the existence of an internally stabilizing controller $C(s)$ for the nominal plant $P(s)$ is equivalent to the existence of a bounded real rational function $S(s)$ that is analytic in $\tilde{\mathbb{C}}_+ := \overline{\mathbb{C}_+} \cup \{\infty\}$, ($\overline{\mathbb{C}_+}$:= closed right half plane), and satisfies the following two interpolation conditions:

(i) The zeros of $S(s)$ contain the poles of $P(s)$ in $\overline{\mathbb{C}_+}$ (multiplicities included).

(ii) The zeros of $S(s)-1$ contain the zeros of $P(s)$ in $\tilde{\mathbb{C}}_+$ (multiplicities included).

Notice it is very important that we consider $\widetilde{\mathbb{C}_+}$ since P may have poles and zeros on the imaginary axis. Moreover, if we take P to be strictly proper, then P will always have a zero at ∞. Thus, these boundary interpolation conditions are necessary to insure not only internal stability but the properness of the compensator C derived from interpolation theory. However for the sake of simplicity we will always assume that the poles and zeros of P in $\widetilde{\mathbb{C}_+}$ are *simple*. In short, the problem of internal stability is one of Lagrange interpolation.

Now let us see why the stability margin optimization problem amounts to one of Nevanlinna-Pick interpolation. Indeed, we have the following:

Lemma 7 *Let $P_k(s) = kP(s)$, $k \in \mathsf{K}$ be as above. Then an internally stabilizing controller $C(s)$ exists for the family of plants $P_k(s)$ if and only if*

$$S(s) := (1 + P(s)C(s))^{-1}$$

satisfies the standard interpolation conditions (i)-(ii), and

$$S(s) : \widetilde{\mathbb{C}_+} \to \mathsf{G} := \mathbb{C}\backslash\{\frac{k}{k-1} \ : \ k \in \mathsf{K}\}. \tag{3.32}$$

Proof. Indeed to have nominal stability $S(s)$ must satisfy (i)-(ii). Moreover, since the poles and zeros of the family P_k are fixed, the same remark holds for the sensitivity functions associated to each member of the family. Now clearly, for each $k \in \mathsf{K}$, we must have,

$$1 + kP(s)C(s) \neq 0 \tag{3.33}$$

for all $s \in \widetilde{\mathbb{C}_+}$. But via some elementary algebraic manipulations, it is easy to see that (3.32) is equivalent to (3.33). □

Now it is elementary to check that G defined above is a simply connected subdomain of $\mathbb{C}$ containing 0 and 1. Hence the generalized stability margin optimization problem amounts to finding (if possible) a real rational analytic function $S(s) : \widetilde{\mathbb{C}_+} \rightarrow \mathsf{G}$ satisfying the interpolation conditions (i) and (ii).

The basic observation is that since G is conformally equivalent to a disc this amounts to a standard problem in Nevanlinna-Pick interpolation. In the next chapter, we will carry out this program.

Chapter 4

One Block Problems

4.1 Optimal stability margin

4.1.1 Generalized problem

We have just seen that the generalized stability margin problem may be formulated as a special case of the following general problem: Given G a simply connected subdomain of $\mathbb{C}$ containing 0 and 1, find (if possible) a real rational analytic function $S(s) : \widetilde{\mathbb{C}_+} \to G$ satisfying the interpolation conditions (i) and (ii) of Section 3.7.

The beautiful fact is that this problem has a solution in terms of $\hat{\gamma}_{opt}$ defined above. More precisely, let $\hat{\gamma}_{opt}(P)$ be the $\hat{\gamma}_{opt}$ relative to the Nevanlinna-Pick problem (see Section 2.11) with the interpolation data

$$\begin{array}{cccccc} a_1 & \dots & a_n & a_{n+1} & \dots & a_{n+m} \\ 1 & \dots & 1 & 0 & \dots & 0 \end{array}$$

where

$$a_i := \frac{z_i - 1}{z_i + 1} \quad i = 1, \dots, n$$

$$a_{n+j} := \frac{p_j - 1}{p_j + 1} \quad j = 1, \ldots, m$$

for $z_i \in \widetilde{\mathbb{C}_+}$ zeros of P, $i = 1, \ldots, n$, and $p_j \in \overline{\mathbb{C}_+}$ poles of P, $j = 1, \ldots, m$. We assume that this interpolation problem has at least one pair of interpolation data, i.e. $n + m \geq 1$.

Now let $\phi : \mathsf{G} \to \mathsf{D}$ be a conformal equivalence such that $\phi(0) = 0$.

Theorem 13 *The generalized stability margin problem is solvable if and only if*

$$|\phi(1)|^{-1} > \hat{\gamma}_{opt}(P).$$

Proof. The stability margin problem is equivalent to finding an analytic function $S : \widetilde{\mathbb{C}_+} \to \mathsf{G}$ which satisfies the interpolation conditions $S(z) = 1$ for each $z \in \widetilde{\mathbb{C}_+}$ a zero of P, and $S(p) = 0$ for each $p \in \overline{\mathbb{C}_+}$ a pole of P. Since ϕ is invertible, and $\phi(0) = 0$, we have that equivalently the stability margin problem is solvable if and only if there exists an analytic function $\widehat{S}(z) := (\phi \circ S)(\frac{1+z}{1-z}) : \overline{\mathsf{D}} \to \mathsf{D}$ which satisfies the interpolation conditions $\widehat{S}(a_i) = \phi(1)$, $i = 1, \ldots, n$, and $\widehat{S}(a_i) = \phi(0) = 0$, $i = n+1, \ldots, n+m$.

If such a function $\widehat{S}$ exists, then $\|\widehat{S}\|_\infty < 1$, and $\widehat{S}/\phi(1)$ is a solution of the Nevanlinna-Pick interpolation problem and therefore

$$\hat{\gamma}_{opt}(P) \leq \|\widehat{S}/\phi(1)\|_\infty < |\phi(1)|^{-1}.$$

Conversely, if $\hat{\gamma}_{opt} < |\phi(1)|^{-1}$ choose $\gamma \in (\hat{\gamma}_{opt}, |\phi(1)|^{-1})$, then there exists an analytic function f, satisfying the interpolation conditions, such that $\|f\|_\infty < \gamma$. Therefore, $\widehat{S} = \phi(1)f$ will satisfy the required conditions. $\square$

Remarks.

i. We should note that $|\phi(1)|$ is independent of the conformal equivalence $\phi : \mathsf{G} \to \mathsf{D}$ with $\phi(0) = 0$. Indeed this follows from the fact that any two such conformal equivalences differ by a factor of $\exp j\theta$ for some $\theta \in [0, 2\pi)$.

ii. One can show from the above proof that the controller corresponding to a given $\widehat{S}$ is

$$C(s) = \frac{1 - \phi^{-1}(\widehat{S}(s))}{\phi^{-1}(\widehat{S}(s))P(s)}$$

which gives an explicit expression for the optimal stabilizing compensator. Notice that we can get all the suboptimal solutions as well from Section 2.11.

In the next section we will apply the theorem to solve the gain and phase margin problems. We will now solve the gain and phase margin problems using Lemma 7 and Theorem 13.

4.1.2 Gain margin optimization

In this case, we have that

$$\mathsf{G} = \mathbb{C} \backslash \{(-\infty, \frac{a}{a-1}] \cup [\frac{b}{b-1}, \infty)\}.$$

The conformal mapping $\phi : \mathsf{G} \to \mathsf{D}$ with $\phi(0) = 0$ is given by:

$$\phi(s) := \frac{1 - [(1 - (\frac{b-1}{b})s)/(1 - (\frac{a-1}{a})s]^{1/2}}{1 + [(1 - (\frac{b-1}{b})s)/(1 - (\frac{a-1}{a})s]^{1/2}}.$$

It is easy to compute that

$$\phi(1) = \frac{1 - \sqrt{a/b}}{1 + \sqrt{a/b}}$$

and thus from Theorem 13, we see that the gain margin problem is solvable iff

$$\frac{1+\sqrt{a/b}}{1-\sqrt{a/b}} > \hat{\gamma}_{opt}(P)$$

or equivalently

$$b/a < \left(\frac{1+\hat{\gamma}_{opt}(P)}{1-\hat{\gamma}_{opt}(P)}\right)^2.$$

Note as $\hat{\gamma}_{opt}(P) \to 1$, the maximal obtainable gain margin goes to ∞. As $\hat{\gamma}_{opt}(P) \to \infty$, the maximal obtainable gain margin goes to zero (in decibels). For minimum-phase stable plants the only interpolation points lie on $\mathbf{T}$, therefore $\hat{\gamma}_{opt}(P) = 1$. For nonminimum phase plants $\hat{\gamma}_{opt}(P) > 1$. See also [96].

4.1.3 Phase margin optimization

In this case we have that

$$\mathsf{G} = \mathbb{C}\backslash\{\frac{s}{s-1} : s = \exp j\theta,\ \theta \in [-\theta_1, \theta_1],\ \theta_1 \in (0, \pi]\}.$$

Using the conformal mapping ϕ from Section 2.10(c) above, it is easy to compute that

$$|\phi(1)| = \sin\frac{\theta_1}{2},$$

and $\phi(0) = 0$. Hence, the phase margin problem is solvable iff

$$\theta < 2\arcsin\hat{\gamma}_{opt}(P)^{-1}.$$

One can check that as $\hat{\gamma}_{opt}(P) \to 1$, the maximal obtainable phase margin (in radians) goes to π. As $\hat{\gamma}_{opt}(P) \to \infty$, the maximal obtainable phase margin goes to zero. See also [16].

This means that $\hat{\gamma}_{opt}(P)$ is a quantity which depends only on the right half zeros and poles of the given plant, and which moreover gives an *exact quantitative measure of its robustness properties.* Large $\hat{\gamma}_{opt}(P)$ means that the given plant will have poor stability margins, and be sensitive to parameter variations, and conversely for small $\hat{\gamma}_{opt}(P)$. This solves both the gain and phase margin problems.

4.2 Interpolation approach

In this section, we apply Nevanlinna-Pick interpolation theory in order to solve the $\mathcal{H}^\infty$ optimal sensitivity and robust stabilization problems for *finite dimensional plants*, with possibly *infinite dimensional weights.* Also discussed in the section is stability margin optimization and unweighted sensitivity minimization for delay systems. Later in this chapter we will solve these problems for distributed plants with rational weighting functions.

4.2.1 Optimal sensitivity

We will first discuss the solution to weighted sensitivity minimization with (stable, proper, real, possibly irrational) weighting filter W_1, we assume $W_1, W_1^{-1} \in \mathcal{H}^\infty$. Indeed, C internally stabilizes the closed loop system, with plant P, if and only if $W_1 S$ is an analytic function in $\widetilde{\mathbb{C}_+}$ such that $(W_1 S)(p) = 0$ for every pole $p \in \overline{\mathbb{C}_+}$ of P, and $(W_1 S)(z) = W_1(z)$ for every zero $z \in \widetilde{\mathbb{C}_+}$ of P. Using Nevanlinna-Pick interpolation, we can easily find the quantity $\hat{\gamma}_{opt}(P, W_1)$ such that

$$\inf\{\|W_1 S\|_\infty : C \text{ internally stabilizing}\} = \hat{\gamma}_{opt}(P, W_1),$$

where, $\hat{\gamma}_{opt}(P, W_1)$ is the $\hat{\gamma}_{opt}$ defined relative to the interpolation data

$$\begin{array}{cccccc} a_1 & \ldots & a_n & a_{n+1} & \ldots & a_{n+m} \\ W_1(z_1) & \ldots & W_1(z_n) & 0 & \ldots & 0 \end{array}$$

with

$$a_i := \frac{z_i - 1}{z_i + 1} \quad i = 1, \ldots, n$$

$$a_{n+j} := \frac{p_j - 1}{p_j + 1} \quad j = 1, \ldots, m$$

for $z_i \in \overline{\mathbb{C}}_+$ zeros of P, and $p_j \in \overline{\mathbb{C}}_+$ poles of P. Note that for $W_1 = 1$, we have that $\hat{\gamma}_{opt}(P, 1) = \hat{\gamma}_{opt}(P)$. Alternatively $\hat{\gamma}_{opt}(P, W_1)$ may be characterized as the smallest $\gamma > 0$ for which there exists an analytic function $f_\gamma : \overline{\mathbb{C}}_+ \to \mathbf{D}$ with $f_\gamma(p) = 0$, and $f_\gamma(z) = \gamma^{-1} W_1(z)$, where p and z are as above. Notice that this type of sensitivity minimization problem will arise as a weighted disturbance attenuation problem (set $W_n = 0$ in Section 3.5). Also, it can be view an asymptotic tracking problem; see [16].

4.2.2 Robust stability

Let us now briefly review the robust stability problem from Chapter 3; see also [61] and [16]. Consider the family of plants defined by

$$(1 + \Delta W_2)P \tag{4.1}$$

where

(i) P and $(1 + \Delta W_2)P$ have the same number of poles in Re $s \geq 0$;

(ii) $\|\Delta\|_\infty \leq \gamma$.

A key result [16] is that there exists an internally stabilizing compensator for the family of plants (4.1) if and only if $\|W_2 T\|_\infty \leq \gamma$, where

$$T = \frac{PC}{1 + PC} = 1 - S$$

is the complementary sensitivity.

Now from this result, it is easy to use Nevanlinna-Pick interpolation theory to compute the maximal γ for which there exists C internally stabilizing for the family (4.1), and moreover explicitly parametrize all such internally stabilizing compensators. Indeed, with the above notation, we have that

$$\inf\{\|W_2T\|_\infty : C \text{ internally stabilizing}\} = \hat{\gamma}_{opt}(P^{-1}, W_2) \qquad (4.2)$$

To see this just note that a point $q \in \mathbb{C}$ is a pole of P if and only if q is a zero of P^{-1}, and similarly $\hat{q} \in \mathbb{C} \cup \{\infty\}$ is a zero of P if and only if $\hat{q}$ is a pole of P^{-1}. That is, in P and P^{-1} the poles and zeros are switched. Thus $\hat{\gamma}_{opt}(P^{-1}, W_2)$ is the smallest γ for which there exists an analytic $f_\gamma : \widetilde{\mathbb{C}_+} \to \mathbb{D}$ with $f_\gamma(p) = \gamma^{-1}W_2(p)$, and $f_\gamma(z) = 0$, for all poles p and zeros z of P.

But now C internally stabilizes the closed loop system if and only if W_2T is an analytic function in $\widetilde{\mathbb{C}_+}$ satisfying the interpolation conditions that $(W_2T)(p) = W_2(p)$ and $(W_2T)(z) = 0$, for all poles p and zeros z of P, from which we get (4.2).

In particular, $\hat{\gamma}_{opt}(P^{-1}, W_2)$ is the maximal tolerance γ such that $\|\Delta\|_\infty \leq \gamma$ for which there exists an internally stabilizing compensator.

4.2.3 Interpolation approach for delay systems

While for the weighted sensitivity minimization problem for even the simplest delay system, we have to use some fairly deep techniques from operator theory (see next sections), it turns out that for gain and phase margin, and unweighted sensitivity optimization for such systems, one can again use elementary Nevanlinna-Pick interpolation theory to solve the given problem. In fact one can even use this technique to solve such problems for the largest class of distributed systems of interest today in control theory ([23]), but we shall suffice here to study the following kind of plant model [59], [36]:

$$P(s) = e^{-hs}P_o(s),$$

where e^{-hs} is the transfer function of a time delay element with the amount of delay $h > 0$, while $P_o(s)$ is a strictly proper real rational function. We shall explicitly solve the unweighted sensitivity minimization problem for $P(s)$, and leave as exercises the corresponding results for gain and phase margin optimization.

Now in order to solve these problems, we shall first have to define the corresponding $\hat{\gamma}_{opt}(P)$ as in Section 2.11.3.

Denote the zeros of $P_o(s)$ in $\overline{\mathbb{C}}_+$ by $z_1, \ldots, z_n$, and the poles by $p_1, \ldots, p_m$. Set

$$a_i := \frac{z_i - 1}{z_i + 1} \quad i = 1, \ldots, n$$

$$a_{j+n} := \frac{p_j - 1}{p_j + 1} \quad j = 1, \ldots, m.$$

Now let $\hat{\gamma}_{opt}(P)$ be the $\hat{\gamma}_{opt}$ defined relative to the interpolation data

$$\begin{matrix} a_1 & \ldots & a_n & a_{n+1} & \ldots & a_{n+m} \\ 0 & \ldots & 0 & e^{hp_1} & \ldots & e^{hp_m} \end{matrix}$$

Let $S := (1 + PC)^{-1}$ denote the sensitivity function. Then we have the following solution to the $\mathcal{H}^\infty$-optimal unweighted sensitivity problem for the plant P:

Theorem 14 *With the above notation,*

$$\inf\{\|S\|_\infty : C \textit{ stabilizing}\} = \hat{\gamma}_{opt}(P).$$

Proof. Again we use the same method of proof as in the rational case. Indeed suppose that $\gamma > 0$ is such that

$$S : \overline{\mathbb{C}}_+ \to \mathsf{D}_\gamma := \{|z| < \gamma\}. \tag{4.3}$$

Since P_o is strictly proper $\gamma > 1$. Clearly, we must compute the infimum over all γ such that there exists an internally stabilizing proper compensator C with (4.3) holding. But (4.3) holds if and only if

$$e^{-hs}P_o(s)C(s) : \widetilde{\mathbb{C}_+} \to \mathsf{G}$$

where $\mathsf{G} := \{z \in \mathbb{C} : |z+1| > 1/\gamma\}$.

Define now the conformal equivalence $\psi : \mathsf{G} \to \mathbf{D}$ by

$$\psi(z) := \frac{\gamma(z+1)-\gamma}{\gamma^2(z+1)-1} = \frac{\gamma z}{\gamma^2(z+1)-1}$$

Notice that $\psi(0) = 0$, and $\psi(\infty) = 1/\gamma$.

Set $u(s) = P_o(s)C(s)$. Then $\psi(e^{-hs}u(s)) = e^{-hs}q(s)$, and since e^{-hs} is inner with no finite zeros, we have that $q(s)$ is analytic on $\widetilde{\mathbb{C}_+}$ and $q : \widetilde{\mathbb{C}_+} \to \mathbf{D}$. Following the same line of reasoning as in the rational case, the interpolation conditions of internal stability translate into the following interpolation conditions on $q(s)$:

$$q(z_i) = 0, \quad i = 1, \ldots, n$$

$$q(p_k) = \frac{e^{hp_k}}{\gamma}, \quad k = 1, \ldots, m.$$

Let now $\hat{\gamma}_{opt}(P)$ be as above. Then from the definition of $\hat{\gamma}_{opt}(P)$, and the above results (see in particular Theorem 9), we have that

$$\gamma > \hat{\gamma}_{opt}(P).$$

Conversely, if $\gamma > \hat{\gamma}_{opt}(P)$ then according to Theorem 9 we can find a rational $q : \widetilde{\mathbb{C}_+} \to \mathbf{D}$ satisfying the above interpolation conditions. Set

$$S_q(s) = \frac{\gamma(1-\gamma e^{-hs}q(s))}{\gamma - e^{-hs}q(s)}$$

since $\gamma > 1 > \|q\|_\infty$, S_q is clearly a bounded analytic function on $\widetilde{\mathbb{C}}_+$ and obviously $|S_q(s)| < \gamma$ on $\widetilde{\mathbb{C}}_+$. Moreover,

$$C(s)S_q(s) = \frac{(\gamma^2-1)q(s)}{\gamma - e^{-hs}q(s)P_o(s)}$$

$$P(s)S_q(s) = e^{-hs}P_o(s)\frac{\gamma(1-\gamma e^{-hs}q(s))}{\gamma - e^{-hs}q(s)}$$

and both of these functions are bounded and analytic in $\widetilde{\mathbb{C}}_+$ because of the interpolation conditions imposed on $q(s)$. Therefore, $C = \frac{1-S}{PS}$ is stabilizing and therefore $\inf\{\|S\|_\infty \ : \ C \text{ stabilizing}\} \leq \|S_q\|_\infty \leq \gamma$, which completes the proof of the theorem. □

4.3 Skew Toeplitz approach

We will see in the next chapter that the two-block $\mathcal{H}^\infty$ control problem defined by (3.29) can be reduced to a one-block problem using a spectral factorization. In the most general form the one-block SISO $\mathcal{H}^\infty$ problem amounts to finding

$$\inf_{q\in\mathcal{H}^\infty} \|\varphi - q\|_\infty \tag{4.4}$$

where $\varphi \in \mathcal{L}^\infty$ is known (obtained from the problem data). The function φ depends on the plant and the weights, and usually it is in the form $\varphi = m^*w$ where m is inner (possibly irrational) and w is rational. In this section we consider the sensitivity minimization problem for stable (possibly infinite dimensional) plants with finite dimensional weighting, and formulate it as a one-block $\mathcal{H}^\infty$ problem. We will use an operator theoretic method, called the *skew Toeplitz approach,* to solve this problem.

Let us first recall the weighted sensitivity minimization problem. The sensitivity function of the closed loop system $[C, P]$ is $S = (1 + PC)^{-1}$. Given a weighting function W_1 we want to find the optimal sensitivity level

$$\gamma_{opt} = \inf_{[C,P]\ stable} \|W_1(1+PC)^{-1}\|_\infty, \tag{4.5}$$

and the optimal controller C_{opt} achieving γ_{opt}. This problem can be obtained by choosing $W_2 = 0$ in (3.29), i.e. we are dealing with the "nominal performance" problem. For simplicity, in this section we will assume that the plant is stable. The general case will be discussed in the following chapters. For technical reasons we will replace $\mathbb{C}_+$ with $\mathbb{D}$, see Section 2.3. We will further assume that the outer part of the plant is invertible. In other words the plant is in the form $p = m_n n_1$, where m_n is arbitrary inner and n_1 is outer and invertible in $\mathcal{H}^\infty(\mathbb{D})$. From the practical control theoretic point of view this problem is not as interesting as the two block problem, but its solution can be used in finding a solution to the two block problem defined in Chapter 3. Throughout this section m_n will be denoted by m.

Choosing $d(z) = 1$, $n(z) = p(z)$, $x(z) = 0$, $y(z) = 1$, we can show that $n, x, d, y \in \mathcal{H}^\infty$ satisfy the Bezout equation (3.3). Then using Theorem 10 we see that all stabilizing controllers are in the form

$$c(z) = \frac{q(z)}{1 - p(z)q(z)}, \quad q \in \mathcal{H}^\infty(\mathbb{D}). \tag{4.6}$$

Substituting (4.6) into (4.5) we have

$$\gamma_{opt} = \inf_{q \in \mathcal{H}^\infty} \|w_1(1 - pq)\|_\infty. \tag{4.7}$$

Since n_1 and w_1 are invertible in $\mathcal{H}^\infty$ (by assumptions, see Section 3.6.1) we can absorb them into the free parameter q. That is, defining $q_1 = w_1 n_1 q$, or $q = q_1 n_1^{-1} w_1^{-1}$, we have

$$\gamma_{opt} = \inf_{q_1 \in \mathcal{H}^\infty} \|w_1 - m q_1\|_\infty. \tag{4.8}$$

This is a one-block $\mathcal{H}^\infty$ control problem of the form (4.4), with $\varphi = m^* w_1$. After finding an optimal q_1^{opt} achieving the optimal performance γ_{opt} we can find the optimal controller from

$$c_{opt}(z) = \frac{q_{opt}(z)}{1 - p(z)q_{opt}(z)} = \frac{q_1^{opt}(z)n_1^{-1}(z)}{w_1(z) - m(z)q_1^{opt}(z)}. \tag{4.9}$$

4.3.1 Nehari's result

The problem (4.8) is equivalent to the Nehari problem, and its solution is given in terms of the norm of the Hankel operator $\mathbf{\Gamma}_{m^*w_1}$ whose symbol is

$$(m^*w_1)(e^{j\theta}) = m(e^{j\theta})^*w_1(e^{j\theta}).$$

The fundamental result by Nehari, [73], is the following.

Theorem 15 (Nehari) *For any $\varphi \in \mathcal{L}^\infty$*

$$\inf_{q\in\mathcal{H}^\infty} \|\varphi - q\|_\infty = \|\mathbf{\Gamma}_\varphi\|.$$

Corollary With the above notation we have

$$\gamma_{opt} = \|\mathbf{\Gamma}_{m^*w_1}\|.$$

Proof. Since m^* is unitary on the unit circle we have

$$\begin{aligned}\gamma_{opt} &= \inf_{q_1\in\mathcal{H}^\infty(\mathsf{D})} \|m^*(w_1 - mq_1)\|_\infty \\ &= \inf_{q_1\in\mathcal{H}^\infty(\mathsf{D})} \|m^*w_1 - q_1\|_\infty.\end{aligned}$$

The last equality means that γ_{opt} is the smallest distance from the $\mathcal{L}^\infty(\mathsf{T})$ function m^*w_1 to the $\mathcal{H}^\infty(\mathsf{D})$ functions; and the result follows from [73]. □

There is an intimate relation between $\mathbf{\Gamma}_{m^*w_1}$ and Sarason's operator $w_1(\mathbf{T})$, where $\mathbf{T}$ is the shift operator compressed to $\mathcal{H}(m)$.

Theorem 16 $\gamma_{opt} = \|\mathbf{\Gamma}_{m^*w_1}\| = \|w_1(\mathbf{T})\|.$

Proof. This result can be deduced by combining Sarason's theorem [89] with Nehari's theorem [73], as follows. Using the above corollary we want to show that $\|\mathbf{\Gamma}_{m^*w_1}\| = \|w_1(\mathbf{T})\|$. Note that when $f \in m\mathcal{H}^2$ we have $m^*w_1 f \in \mathcal{H}^2$. So $\mathbf{\Gamma}_{m^*w_1} f = 0$ for all $f \in m\mathcal{H}^2$. Hence

$$\|\mathbf{\Gamma}_{m^*w_1}\| = \|\mathbf{\Gamma}_{m^*w_1}|_{\mathcal{H}(m)}\|.$$

We claim that $\mathbf{\Gamma}_{m^*w_1}|_{\mathcal{H}(m)} = m^*w_1(\mathbf{T})$. Assume this is true for a moment, then we have, since m^* is unitary,

$$\|\mathbf{\Gamma}_{m^*w_1}|_{\mathcal{H}(m)}\| = \|m^*w_1(\mathbf{T})\| = \|w_1(\mathbf{T})\|$$

and this completes the proof.

Now we prove the above claim. Take any $f \in \mathcal{H}(m)$ and define $w_1(\mathbf{T})f =: g$. But

$$\begin{aligned} g &= \mathbf{P}_{\mathcal{H}(m)} w_1 f = w_1 f - m\mathbf{P}_+ m^* w_1 f, \\ m^* g &= m^* w_1 f - \mathbf{P}_+ m^* w_1 f, \\ \mathbf{P}_- m^* g &= \mathbf{P}_- m^* w_1 f. \end{aligned}$$

Since $g = w_1(\mathbf{T})f \in \mathcal{H}(m)$, we have $m^*g \in \mathcal{L}^2 \ominus \mathcal{H}^2$. Thus, for any $f \in \mathcal{H}(m)$ we have

$$m^*w_1(\mathbf{T})f = m^*g = \mathbf{P}_- m^*g = \mathbf{P}_- m^* w_1 f = \mathbf{\Gamma}_{m^*w_1} f$$

as claimed. □

In order to compute the $\mathcal{H}^\infty$ optimal performance level γ_{opt} we want to find the norm of $w_1(\mathbf{T})$. Since the operator $w_1(\mathbf{T})$ is of infinite rank, its norm is the largest of two quantities: the essential norm, denoted by $\|w_1(\mathbf{T})\|_e$ (see Section 2.4.1), and the largest singular value. For the operator $w_1(\mathbf{T})$ the essential norm can be computed as (see e.g. [33], [34], [81], etc.)

$$\|w_1(\mathbf{T})\|_e = \max_\theta \{\ |w_1(e^{j\theta})|\}. \tag{4.10}$$

where $e^{j\theta}$ is an essential singularity of $m(z)$. Note that when m has finitely many essential singularities it is trivial to compute the essential norm: we simply have to evaluate w_1 at finitely many points.

The identity (4.10) gives a lower bound for γ_{opt}. On the other hand if we choose $q_1 = 0$ in (4.8) we get an upper bound for γ_{opt}: $\|w_1\|_\infty$. Let us assume $\|w_1(\mathbf{T})\| > \|w_1(\mathbf{T})\|_e$ so that the norm is achieved at the largest singular value. Then, in order to compute γ_{opt}, we need to find the largest singular value of $w_1(\mathbf{T})$ between $\|w_1(\mathbf{T})\|_e$ and $\|w_1\|_\infty$.

Now we are going to study the necessary and sufficient conditions for a given number $\rho \in (\|w_1(\mathbf{T})\|_e, \|w_1\|_\infty)$ to be a singular value, i.e. existence of a non-zero singular vector for ρ. The singular value/singular vector equation for $w_1(\mathbf{T})$ is

$$\left(\rho^2\mathbf{I} - w_1(\mathbf{T})^* w_1(\mathbf{T})\right) y = 0. \tag{4.11}$$

Recall that the weight w_1 we consider is rational and $w_1 \in \mathcal{H}^\infty$. So we can write $w_1(z) = b(z)/k(z)$ where $b(z)$ and $k(z)$ are polynomials with $1/k \in \mathcal{H}^\infty$, in particular $k(\mathbf{T})^{-1} = \frac{1}{k}(\mathbf{T})$ exists. Let n be the maximum of degrees of $b(z)$ and $k(z)$, i.e.

$$\begin{aligned} b(z) &= b_0 + b_1 z^1 + \cdots + b_n z^n \\ k(z) &= k_0 + k_1 z^1 + \cdots + k_n z^n \end{aligned}$$

k_n or b_n is nonzero by definition. With this notation we have $w_1(\mathbf{T}) = b(\mathbf{T})k(\mathbf{T})^{-1}$. Now define $k(\mathbf{T})^{-1}y =: u$, since $1/k \in \mathcal{H}^\infty$ and $\|\mathbf{T}\| \leq 1$, $u \in \mathcal{H}(m)$ if and only if $y \in \mathcal{H}(m)$. Thus equation (4.11) can be multiplied by $k(\mathbf{T})^*$ on the left and expressed in terms of u.

This easily leads to the following.

Lemma 8 *Assume that $\gamma_{opt} > \|w_1(\mathbf{T})\|_e$. Then γ_{opt} is the largest value of ρ for which there is a non-zero $u \in \mathcal{H}(m)$ satisfying*

$$\left(b(\mathbf{T})^* b(\mathbf{T}) - \rho^2 k(\mathbf{T})^* k(\mathbf{T})\right) u = 0. \tag{4.12}$$

4.3.2 Skew Toeplitz operators

Note that (4.12) is in the form $\mathbf{A}_\rho u = 0$, where

$$\mathbf{A}_\rho := b(\mathbf{T})^* b(\mathbf{T}) - \rho^2 k(\mathbf{T})^* k(\mathbf{T}).$$

Operators of the form $\mathbf{A}_\rho$ are called *skew Toeplitz*, [6]. Conditions on the invertibility of this skew Toeplitz operator determines the $\mathcal{H}^\infty$ optimal performance, γ_{opt}. Note that since b and k are polynomials in z, $\mathbf{A}_\rho$ is in the form

$$\mathbf{A}_\rho =: \sum_{i,j=0}^{n} c_{ij}\mathbf{T}^{*i}\mathbf{T}^{j} \quad \text{where} \quad c_{ij} = c_{ji}^{*} \in \mathbb{C}. \tag{4.13}$$

So $\mathbf{A}_\rho$ is a *polynomial* in powers of $\mathbf{T}^*$ and $\mathbf{T}$. As mentioned in Chapter 2, when m is rational, $\mathbf{T}$ is finite dimensional (i.e. a square matrix of finite size), and in this case the skew Toeplitz operator $\mathbf{A}_\rho$ is a finite size square matrix and we can easily determine conditions on its invertibility. However, when m is an arbitrary inner function $\mathbf{A}_\rho$ is infinite dimensional. Although most of our further discussion is also valid for functions which are not real, in control theory, as already mentioned in Section 3.6, they are assumed to be real. In particular this means that the coefficients of b and k are real, and therefore so are the coefficients c_{ij}'s in (4.13). Also since m is real we have $m(\zeta)^* = \overline{m(\zeta)} = m(1/\zeta)$ for $\zeta \in \mathbb{T}$.

4.3.3 On the optimal controller

We have already seen that the largest singular value of $w_1(\mathbf{T})$ determines the optimal performance level γ_{opt}. The optimal controller can be found from q_1^{opt} as follows. It is easy to verify that $w_1(\mathbf{T})\mathbf{T} = \mathbf{T}w_1(\mathbf{T})$, i.e. the operator $w_1(\mathbf{T})$ commutes with the compressed shift. Therefore, by Sarason's theorem there exists a minimal dilation $s_{opt} \in \mathcal{H}^\infty$ such that $s_{opt}(\mathbf{T}) = w_1(\mathbf{T})$ and $\|s_{opt}\|_\infty = \|w_1(\mathbf{T})\| = \gamma_{opt}$. Since $s_{opt}(\mathbf{T}) = w_1(\mathbf{T})$, by remark (iv) in Section 2.7, we have $s_{opt} = w_1 - mq_1^{opt}$, for some $q_1^{opt} \in \mathcal{H}^\infty$.

We can obtain s_{opt} (and hence q_1^{opt} and c_{opt}) from the optimal performance γ_{opt}, i.e. the largest singular value of $w_1(\mathbf{T})$, and a singular vector $y_{opt} = k(\mathbf{T})u_{opt}$ corresponding to γ_{opt} satisfying the equivalent singular value/singular vector equation (4.12):

$$\left(b(\mathbf{T})^*b(\mathbf{T}) - \gamma_{opt}{}^2 k(\mathbf{T})^*k(\mathbf{T})\right) u_{opt} = 0.$$

Theorem 17 *The minimal dilation* $s_{opt}(z) = w_1(z) - m(z)q_1^{opt}(z)$ *of* $w_1(\mathbf{T})$ *is given by*

$$s_{opt} = \frac{w_1(\mathbf{T})y_{opt}}{y_{opt}} = \frac{b(\mathbf{T})u_{opt}}{k(\mathbf{T})u_{opt}}. \tag{4.14}$$

Proof. This theorem is the same as Theorem 6 (with slight modifications in the notation), and the result follows from Proposition 5.1 of [89]. For convenience to the reader, we give the complete proof here. The second equality is obvious from the definition of u_{opt}, y_{opt} and $w_1 = b/k$. For the first equality we need to show that

$$\begin{aligned} s_{opt}y_{opt} &= w_1(\mathbf{T})y_{opt} = s_{opt}(\mathbf{T})y_{opt} \\ &= \mathbf{P}_{\mathcal{H}(m)}s_{opt}y_{opt}. \end{aligned} \tag{4.15}$$

In other words we need to show that $s_{opt}y_{opt}$ lies in $\mathcal{H}(m)$, which is equivalent to

$$\|s_{opt}y_{opt}\| = \|\mathbf{P}_{\mathcal{H}(m)}s_{opt}y_{opt}\|. \tag{4.16}$$

Let us define

$$\widehat{y}_{opt} = \gamma_{opt}^{-1}w_1(\mathbf{T})y_{opt}.$$

Then from (4.11), with $\rho = \gamma_{opt}$, we have

$$y_{opt} = \gamma_{opt}^{-1}w_1(\mathbf{T})^*\widehat{y}_{opt},$$

and the following inequalities hold

$$\begin{aligned} \|w_1(\mathbf{T})^*\|\|\widehat{y}_{opt}\| &= \|w_1(\mathbf{T})^*\|\|\gamma_{opt}^{-1}w_1(\mathbf{T})y_{opt}\| = \|w_1(\mathbf{T})y_{opt}\| \\ &= \|s_{opt}(\mathbf{T})y_{opt}\| = \|\mathbf{P}_{H(m)}s_{opt}y_{opt}\| \\ &\le \|s_{opt}y_{opt}\| \\ &\le \|s_{opt}\|\|y_{opt}\| = \gamma_{opt}\|\gamma_{opt}^{-1}w_1(\mathbf{T})^*\widehat{y}_{opt}\| \\ &\le \|w_1(\mathbf{T})^*\|\|\widehat{y}_{opt}\|. \end{aligned}$$

Thus all above inequalities are in fact equalities, and in particular (4.16) holds. This completes the proof. □

4.3.4 Computation of γ_{opt}

In this section we will derive finitely many linear equations (the so called *singular system*) for computing γ_{opt} and s_{opt} (hence c_{opt}). Our starting point is Lemma 8. We want to examine the conditions under which there exists a non-zero $u \in \mathcal{H}(m)$ satisfying the singular value/singular vector equation (4.12):

$$\left(b(\mathbf{T})^*b(\mathbf{T}) - \rho^2 k(\mathbf{T})^*k(\mathbf{T})\right) u = \mathbf{A}_\rho u = 0.$$

We are going to write the left hand side explicitly, and this will give us the necessary and sufficient conditions on ρ for the existence of a non-zero $u \in \mathcal{H}(m)$. Before going into details we would like to present the main idea behind the computations below.

First, recall that the skew Toeplitz operator $\mathbf{A}_\rho$ is a polynomial in $\mathbf{T}^j$ and $\mathbf{T}^{*j}$, $j = 1, 2, \ldots, n$. It was shown in Chapter 2 that applying $\mathbf{T}$ to an element $u \in \mathcal{H}(m)$ one gets

$$(\mathbf{T}u)(z) = zu(z) - m(z)\phi_{-1},$$

where ϕ_{-1} is obtained from the expansion

$$m^*u = \phi_{-1}z^{-1} + \phi_{-2}z^{-2} + \ldots. \tag{4.17}$$

Again from Chapter 2 we have

$$(\mathbf{T}^*u)(z) = z^{-1}u(z) - z^{-1}\phi_0$$

where ϕ_0 comes from the expansion of u:

$$u(z) = \phi_0 + \phi_1 z + \phi_2 z^2 + \ldots. \tag{4.18}$$

Since $\mathbf{A}_\rho$ is a polynomial, applying $\mathbf{T}^{*j}$ and $\mathbf{T}^j$, $j = 1, 2, \ldots, n$, to $u \in \mathcal{H}(m)$ recursively we get a polynomial (in z and z^{-1} up to powers

n), which multiplies $u(z)$, and additional terms involving $\phi_0, \ldots, \phi_{n-1}$ and $\phi_{-1}, \ldots, \phi_{-n}$. That is, we can show that $\mathbf{A}_\rho u$ is of the form

$$\begin{aligned}(\mathbf{A}_\rho u)(z) &= \left(b(z^{-1})b(z) - \rho^2 k(z^{-1})k(z)\right) u(z) \\ &\quad -[r_1(z)\ r_2(z)\ \ldots\ r_{2n}(z)]\ \phi,\end{aligned}$$

where $\phi = [\,\phi^*_{-n}\ \ \cdots\ \ \phi^*_{-1}\ \ \phi^*_0\ \ \cdots\ \ \phi^*_{n-1}\,]^*$, and $r_j(z), j = 1, \ldots 2n$, are explicitly computable functions, depending on the parameter ρ. Since for ρ to be a singular value there has to be a non-zero $u \in \mathcal{H}(m)$ satisfying $\mathbf{A}_\rho u = 0$, we must have

$$u(z) = \frac{R_\rho(z)\phi}{b(z^{-1})b(z) - \rho^2 k(z^{-1})k(z)} \tag{4.19}$$

where $R_\rho(z)$ is a $1 \times 2n$ vector of functions and ϕ is a $2n \times 1$ constant vector. But the denominator of (4.19) vanishes at its $2n$ roots $z_1, \ldots, z_{2n}$. So, for u to be a non-zero element of $\mathcal{H}(m)$ we must have a non-zero $\phi \in \mathbb{C}^{2n}$ such that (assuming z_j's are distinct)

$$R_\rho(z_j)\phi = 0, \quad \forall\, j = 1, \ldots, 2n.$$

We have $2n$ equations in $2n$ unknowns, which means that there is a non-zero solution if and only if the $2n \times 2n$ complex matrix

$$\begin{bmatrix} R_\rho(z_1) \\ \vdots \\ R_\rho(z_{2n}) \end{bmatrix}$$

is singular. This gives a rank type (or determinantal) formula for ρ to be a singular value of $w_1(\mathbf{T})$. In the discussion below we will derive an explicit formula for $R_\rho(z)$. Then by carefully studying the resulting $2n$ system of linear equations we will be able to reduce the number of equations to n.

We now study the action of each term of $\mathbf{A}_\rho$ on $u \in \mathcal{H}(m)$. Note that we have

$$b(\mathbf{T})u = b(\mathbf{S})u - m(\mathbf{S})\mathbf{P}_+ m^* b u.$$

Since $b(z)$ is an n-th order polynomial and m^*u has an expansion of the form (4.17) the projection $\mathbf{P}_+m^*bu$ is a polynomial of degree $n-1$. It is easy to check that this projection is given by

$$\mathbf{P}_+m^*bu = V_+(z)\mathcal{B}\phi_- \tag{4.20}$$

where $V_+(z) := [1 \quad z \quad \dots \quad z^{n-1}]$,

$$\mathcal{B} := \begin{bmatrix} b_n & \dots & b_1 \\ 0 & \ddots & \vdots \\ 0 & 0 & b_n \end{bmatrix}, \text{ and } \quad \phi_- := \begin{bmatrix} \phi_{-n} \\ \vdots \\ \phi_{-1} \end{bmatrix}.$$

Similarly,

$$k(\mathbf{T})u = k(\mathbf{S})u - m(\mathbf{S})(V_+(z)\mathcal{K}\phi_-)\ , \text{ where } \quad \mathcal{K} := \begin{bmatrix} k_n & \dots & k_1 \\ 0 & \ddots & \vdots \\ 0 & 0 & k_n \end{bmatrix}.$$

Since the operator $\mathbf{T}^*$ is the same as $\mathbf{S}^*$, we can rewrite (4.12) as follows

$$\begin{aligned}\left(b(\mathbf{S})^*b(\mathbf{S}) - \rho^2 k(\mathbf{S})^*k(\mathbf{S})\right)u \quad &- \quad b(\mathbf{S})^*(m(z)V_+(z)\mathcal{B}\phi_-) \\ &- \quad \rho^2 k(\mathbf{S})^*(m(z)V_+(z)\mathcal{K}\phi_-) = 0\end{aligned}$$

Recalling the action of $\mathbf{S}^*$ from Chapter 2, we have

$$(b(\mathbf{S})^*(m(z)V_+(z))\ )\ (z) = V_+(z)m(z)b(z^{-1}) - V_-(z)\mathcal{B}^*\mathcal{M}, \tag{4.21}$$

where, $\mathcal{B}^*$ is the transpose of $\mathcal{B}$,

$$V_-(z) := [z^{-n} \quad \dots \quad z^{-1}], \qquad \mathcal{M} := \begin{bmatrix} m_0 & 0 & 0 \\ \vdots & \ddots & 0 \\ m_{n-1} & \cdots & m_0 \end{bmatrix},$$

and $m(z) =: m_0 + m_1 z + m_2 z^2 + \cdots$. Now define a $2n$-th order polynomial

$$\chi_\rho(z) := z^n(b(z^{-1})b(z) - \rho^2 k(z^{-1})k(z)).$$

Note that $\chi_\rho(z)$ is of the form $\chi_\rho(z) = z^n(\chi_{-n}z^{-n}+\ldots+\chi_0+\ldots+\chi_n z^n)$ with $\chi_i = \chi_{-i}$. Further define

$$\mathcal{X} := \begin{bmatrix} \chi_{-n} & 0 & 0 \\ \vdots & \ddots & 0 \\ \chi_{-1} & \cdots & \chi_{-n} \end{bmatrix}, \text{ and } \phi_+ := \begin{bmatrix} \phi_0 \\ \vdots \\ \phi_{n-1} \end{bmatrix},$$

where $\phi_0, \ldots, \phi_{n-1}$ are as in (4.18). Then we have

$$\left(b(\mathbf{S})^* b(\mathbf{S}) - \rho^2 k(\mathbf{S})^* k(\mathbf{S})\right) u = z^{-n}\chi_\rho(z)u(z) - V_-(z)\mathcal{X}\phi_+.$$

Finally defining $\mathcal{L} := \mathcal{B}^*\mathcal{M}\mathcal{B} - \rho^2\mathcal{K}^*\mathcal{M}\mathcal{K}$, and $\tilde{b}(z) := z^n b(z^{-1})$, (similarly for $\tilde{k}(z)$), we see that (4.12) is equivalent to

$$\chi_\rho(z)u(z) = R_-(z)\phi_- + R_+(z)\phi_+, \tag{4.22}$$

where

$$R_-(z) := V_+(z)\left(m(z)(\tilde{b}(z)\mathcal{B} - \rho^2\tilde{k}(z)\mathcal{K}) - \mathcal{L}\right), \tag{4.23}$$

$$R_+(z) := V_+(z)\mathcal{X}. \tag{4.24}$$

In order to define $R_-(z)$ for z outside $\overline{\mathbf{D}}$ we use the definition $m(z) = 1/m(1/z)$ for m analytic on $\mathbf{D}$. Thus $m(z)$ is analytic outside $\overline{\mathbf{D}}$ with the exception of the points $1/z_o$, where $z_o \in \mathbf{D}$ is a zero of $m(z)$.

This is the explicit form of the entries of the matrix $R_\rho(z)$, which gives us $2n$ equations for u to be a non-zero element of $\mathcal{H}(m)$. The equations are derived as follows.

First we make the following assumption for simplicity.

Assumption 4.1: The roots of $\chi_\rho(z)$ are all non-zero and distinct.

This assumption holds generically and can be relaxed easily, see [38]. We need another assumption which also holds generically, and can be relaxed, [51].

Assumption 4.2: If ζ is a root of χ_ρ then $m(\zeta) \neq 0$.

Let us enumerate the roots of χ_ρ as $z_1, z_2, \ldots, z_{2n}$ in such a way that first r of them are inside the closed unit disc $\overline{\mathbf{D}}$, and the rest are outside. Note that by symmetry if ζ is a root of χ_ρ then so is $1/\zeta$. Therefore, we can order the roots in such a way that $z_{n+i} = 1/z_i$, $i = 1, 2, \ldots, n$. We must now refer to a deeper fact concerning the elements of $\mathcal{H}(m)$. Namely, if $u \in \mathcal{H}(m)$ then u is analytic across those arcs of $\mathbf{T}$ on which $m(z)$ is analytic, see Lemma 2.3, p.355 of [28]. In our case if some $z_i \in \mathbf{T}$, then $|w_1(z_i)|^2 = \rho^2 > \sup\{|W(\zeta)| \ : \ \zeta \in \mathbf{T},\ \zeta \text{ is a singularity of } m\}$ and therefore $u(z)$ is analytic in the neighborhood of z_i. So we must have

$$R_-(z_i)\phi_- + R_+(z_i)\phi_+ = 0 \quad i = 1, \ldots r. \tag{4.25}$$

Also, since $m^* u$ is analytic outside $\overline{\mathbf{D}}$, and according to Assumption 4.2 since $R_-(z)$ is analytic at z_{r+k} for $k = 1, \ldots, 2n - r$, we must also have

$$R_-(z_i)\phi_- + R_+(z_i)\phi_+ = 0 \quad i = r+1, \ldots 2n. \tag{4.26}$$

The following result, which gives the $2n$—equation type of determinantal formula, was obtained in [38]. Similar results appear in [30, 31, 32, 38].

Theorem 18 *([38]) Suppose that Assumptions 4.1 and 4.2 hold and let $\|w_1(\mathbf{T})\|_e < \rho < \|w_1\|_\infty$. Then, ρ is a singular value of $w_1(\mathbf{T})$ if and only if there exists a non-zero $\begin{bmatrix} \phi_- \\ \phi_+ \end{bmatrix} \in \mathbb{C}^{2n}$ which satisfies $2n$ equations given by (4.25) and (4.26):*

$$\begin{bmatrix} R_-(z_1) & R_+(z_1) \\ \vdots & \vdots \\ R_-(z_{2n}) & R_+(z_{2n}) \end{bmatrix} \begin{bmatrix} \phi_- \\ \phi_+ \end{bmatrix} = 0. \tag{4.27}$$

Proof. (See also [28], [38], [35], [76], [78], [79]) The necessity of (4.27) is obvious from the derivation of these equations. The fact that $[\phi_-^T \ \ \phi_+^T]$

is non-zero follows from (4.22), because otherwise $u = 0$. For sufficiency part suppose that there exists a non-zero $\hat{\phi} = \begin{bmatrix} \hat{\phi}_- \\ \hat{\phi}_+ \end{bmatrix} \in \mathbb{C}^{2n}$, such that

$$\begin{bmatrix} R_-(z_1) & R_+(z_1) \\ \vdots & \vdots \\ R_-(z_{2n}) & R_+(z_{2n}) \end{bmatrix} \hat{\phi} = 0.$$

Then it is possible to find a non-zero $\hat{u} \in \mathcal{H}(m)$ from equation (4.22):

$$\hat{u}(z) := \frac{R_-(z)\hat{\phi}_- + R_+(z)\hat{\phi}_+}{\chi_\rho(z)}, \text{ for } z \in \overline{\mathbf{D}}, z \neq z_1, \ldots, z_r$$

Now consider

$$(m^*\hat{u})(z) = m(\frac{1}{z})\hat{u}(z) = \frac{m(\frac{1}{z})R_-(z)\hat{\phi}_- + m(\frac{1}{z})R_+(z)\hat{\phi}_+}{\chi_\rho(z)}$$

is analytic outside $\overline{\mathbf{D}}$ with the possible exception of the zeros of $\chi_\rho(z)$. However, the last $2n - r$ equations in (4.27) impose the cancellations necessary to make $m^*\hat{u}$ analytic at those zeros too. Finally $(m^*\hat{u})(z) \to 0$ for $|z| \to \infty$ since $|m(1/z)| \leq 1$ outside $\overline{\mathbf{D}}$ and the denominator is of higher degree than the polynomials appearing in the numerator. Thus $m^*\hat{u}$ is analytic outside $\overline{\mathbf{D}}$ (and at ∞) and therefore $\hat{u} \in \mathcal{H}(m)$.

We now want to show that the vector ϕ defined from $\hat{u}$ is precisely $\hat{\phi}$, so that the singular value/singular vector equation is consistent. That is the vector ϕ for $\hat{u}$, which is given by the coefficients of

$$\begin{aligned} \hat{u}(z) &= \phi_0 + \phi_1 z^1 + \phi_2 z^2 + \cdots, \\ (m^*\hat{u})(z) &= \phi_{-1} z^{-1} + \phi_{-2} z^{-2} + \cdots, \end{aligned}$$

must be the same as $\hat{\phi}$. Note that $\hat{u}$ satisfies

$$\hat{u} = \frac{b(\mathbf{S})^*(mV_+)\mathcal{B}\hat{\phi}_- - \rho^2 k(\mathbf{S})^*(mV_+)\mathcal{K}\hat{\phi}_- + V_-\mathcal{X}\hat{\phi}_+}{b(z^{-1})b(z) - \rho^2 k(z^{-1})k(z)}. \qquad (4.28)$$

Multiplying $\hat{u}(z)$ by the denominator of the right hand side of the above equation, and taking the orthogonal projection onto $\mathcal{L}^2 \ominus \mathcal{H}^2$ we obtain

$$V_-(z)\mathcal{X}\phi_+ = V_-(z)\mathcal{X}\hat{\phi}_+.$$

Since the entries of $V_-(z)$ span an $n-$dimensional space and $\mathcal{X}$ is non-singular (because Assumption 4.1 implies that $\chi_\rho(0) \neq 0$, and this means $\chi_{-n} \neq 0$ hence $\mathcal{X}$ is invertible) we have that $\hat{\phi}_+ = \phi_+$. Also multiplying $\hat{u}$ by the denominator of (4.28) and m^*, and then taking the orthogonal projection on $\mathcal{H}^2$ we get

$$\begin{aligned} &b(\mathbf{S})^* m(\mathbf{S})^* b(\mathbf{S})\hat{u} - \rho^2 k(\mathbf{S})^* m(\mathbf{S})^* k(\mathbf{S})\hat{u} \\ &\quad = b(\mathbf{S})^*(V_+(z)\mathcal{B}\hat{\phi}_-) - \rho^2 k(\mathbf{S})^*(V_+(z)\mathcal{K}\hat{\phi}_-), \end{aligned} \tag{4.29}$$

where we have used

$$\mathbf{P}_+ m^* b(z^{-1})b(z) = m(\mathbf{S})^* b(\mathbf{S})^* b(\mathbf{S}) = b(\mathbf{S})^* m(\mathbf{S})^* b(\mathbf{S}).$$

Note also that

$$m(\mathbf{S})^* b(\mathbf{S})\hat{u} = \mathbf{P}_+ m^* b\hat{u} = V_+(z)\mathcal{B}\phi_-.$$

Hence we conclude from (4.29) that

$$\begin{aligned} b(\mathbf{S})^*(V_+(z)\mathcal{B}\phi_-) &- \rho^2 k(\mathbf{S})^*(V_+(z)\mathcal{K}\phi_-) \\ &= b(\mathbf{S})^*(V_+(z)\mathcal{B}\hat{\phi}_-) - \rho^2 k(\mathbf{S})^*(V_+(z)\mathcal{K}\hat{\phi}_-). \end{aligned}$$

This can be re-written as

$$V_+(z)(\hat{\mathcal{B}}\mathcal{B} - \rho^2\hat{\mathcal{K}}\mathcal{K})(\phi_- - \hat{\phi}_-) = 0,$$

where

$$\hat{\mathcal{B}} = \begin{bmatrix} b_0 & \dots & b_{(n-1)} \\ 0 & \ddots & \vdots \\ 0 & 0 & b_0 \end{bmatrix}, \quad \text{and} \quad \hat{\mathcal{K}} = \begin{bmatrix} k_0 & \dots & k_{(n-1)} \\ 0 & \ddots & \vdots \\ 0 & 0 & k_0 \end{bmatrix}.$$

The entries of $V_+(z)$ span an n–dimensional space. Furthermore, $\widehat{\mathcal{B}}\mathcal{B} - \rho^2\widehat{\mathcal{K}}\mathcal{K}$ is an upper triangular matrix whose diagonal entries are $b_0 b_n - \rho^2 k_0 k_n$ which is equal to χ_n (non-zero by Assumption 4.1). Thus, we conclude that $\phi_- = \hat{\phi}_-$, so $\hat{u} \in \mathcal{H}(m)$ defined above is a singular vector for the singular value ρ. □

Now we will make some simplifications in the above $2n$ equations, and show that (4.27) can be reduced to n equations. Defining $F(z) := m(z)(\tilde{b}(z)\mathcal{B} - \rho^2\tilde{k}(z)\mathcal{K})$, (4.27) can be re-written as

$$\begin{bmatrix} V_+(z_1)(F(z_1) - \mathcal{L}) & V_+(z_1)\mathcal{X} \\ \vdots & \vdots \\ V_+(z_{2n})(F(z_{2n}) - \mathcal{L}) & V_+(z_{2n})\mathcal{X} \end{bmatrix} \begin{bmatrix} \phi_- \\ \phi_+ \end{bmatrix} = 0. \tag{4.30}$$

Introducing the Vandermonde matrices

$$\mathcal{V}_+ := \begin{bmatrix} V_+(z_1) \\ \vdots \\ V_+(z_n) \end{bmatrix} \quad \mathcal{V}_- := \begin{bmatrix} V_+(z_1^{-1}) \\ \vdots \\ V_+(z_n^{-1}) \end{bmatrix}$$

and defining

$$\mathcal{F}_+ := \begin{bmatrix} V_+(z_1)F(z_1) \\ \vdots \\ V_+(z_n)F(z_n) \end{bmatrix} \quad \mathcal{F}_- := \begin{bmatrix} V_+(z_1^{-1})F(z_1^{-1}) \\ \vdots \\ V_+(z_n^{-1})F(z_n^{-1}) \end{bmatrix}$$

with $\phi'_+ := \mathcal{V}_+\mathcal{X}\phi_+$, (4.30) becomes

$$\begin{bmatrix} \mathcal{E}_{11} & \mathrm{I} \\ \mathcal{E}_{21} & \mathcal{V}_-\mathcal{V}_+^{-1} \end{bmatrix} \begin{bmatrix} \phi_- \\ \phi'_+ \end{bmatrix} = 0 \tag{4.31}$$

where $\mathcal{E}_{11} := \mathcal{F}_+ - \mathcal{V}_+\mathcal{L}$, and $\mathcal{E}_{21} := \mathcal{F}_- - \mathcal{V}_-\mathcal{L}$. Now eliminating $\phi'_+ = -\mathcal{E}_{11}\phi_-$ from the first n equations of (4.31) the second set of n equations in (4.31) becomes

$$\left(\mathcal{E}_{21} - \mathcal{V}_-\mathcal{V}_+^{-1}\mathcal{E}_{11}\right)\phi_- = 0. \tag{4.32}$$

We can further simplify (4.32) to obtain

$$\left(\mathcal{V}_-^{-1}\mathcal{F}_- - \mathcal{V}_+^{-1}\mathcal{F}_+\right)\phi_- = 0.$$

The above formula appears in [75] and it can be summarized as follows:

Theorem 19 *([75]) Under the assumptions of Theorem 18, γ_{opt} is the largest value of ρ for which there exists a non-zero $\phi_- \in \mathbb{C}^n$ such that*

$$\mathcal{R}_\rho\phi_- = 0,$$

where

$$\mathcal{R}_\rho := \mathcal{V}_-^{-1}\mathcal{F}_- - \mathcal{V}_+^{-1}\mathcal{F}_+.$$

Proof. The result follows from Theorem 18 and the fact that the matrices $\mathcal{V}_+$ and $\mathcal{X}$ (used in the transformation from ϕ_+ to ϕ'_+) are invertible. Invertibility of the Vandermonde matrix $\mathcal{V}_+$ is due to Assumption 4.1. Also by the same assumption $\mathcal{X}$ is invertible. □

Note that all the singular values (not just the norm) of $w_1(\mathbf{T})$ (and of $\mathbf{\Gamma}_{m^*w_1}$) are given by the values of ρ which makes $\mathcal{R}_\rho$ singular. In order to construct $\mathcal{R}_\rho$ we first need to find the roots of $\chi_\rho(z) = 0$. But when n is large it is not possible to compute the roots as explicit functions of ρ. On the other hand we can compute these roots and hence $\mathcal{R}_\rho$ numerically for each fixed value of ρ. Therefore, we can search for γ_{opt} by decreasing ρ from an upper bound e.g. $\|w_1\|_\infty$. At each step we check whether the matrix $\mathcal{R}_\rho$ is singular (or "close" to being singular). This can be done by computing the smallest singular value $\sigma_{min}(\mathcal{R}_\rho)$. Then the zeros in the plot of $\sigma_{min}(\mathcal{R}_\rho)$ versus ρ indicate the location of the singular values of $w_1(\mathbf{T})$, largest of which is the norm, i.e. γ_{opt}. We will illustrate this point via an example in Section 4.3.6.

Another interesting point to remark here is that z_i is a zero of $\chi_\rho(z)$ if and only if it is a pole of $(\rho^2 - w_1(z^{-1})w_1(z))^{-1}$. Therefore, $z_i = \frac{s_i-1}{s_i+1}$ where s_i is a pole of the transfer function (defined in the s−plane)

$$(\rho^2 - W_1(-s)W_1(s))^{-1}, \quad \text{where} \quad W_1(s) = w_1(\frac{s-1}{s+1}).$$

Let $[A,B,C,d]$ be a minimal realization of the function $W_1(s)$, i.e. $W_1(s) = d + C(sI-A)^{-1}B$, where the dimension of A is $n\times n$. Then, one can easily prove that s_i's are given by the eigenvalues of the Hamiltonian matrix

$$H_\rho = \begin{bmatrix} (A+\frac{BdC}{\rho^2}) & -\frac{BB^*}{\rho^2} \\ C^*(I+\frac{d^2}{\rho^2})C & -(A+\frac{BdC}{\rho^2})^* \end{bmatrix}.$$

This formula for the roots of $\chi_\rho(z)$ suggests that there may be interesting connections between the set of n-equations given above and the same number of equations obtained in [65], [90] and [122]. In each of these references the set of $n-$equations are obtained in terms of certain Hamiltonian matrices constructed from the state space realizations of $W_1(s)$, like H_ρ.

4.3.5 Optimal controller

After obtaining the optimal performance level γ_{opt} from the plot of $\sigma_{min}(\mathcal{R}_\rho)$ versus ρ we can obtain the optimal controller from a non-zero ϕ_-^{opt} satisfying

$$\mathcal{R}_{\gamma_{opt}}\phi_-^{opt} = 0.$$

Note that from equation (4.31) ϕ_-^{opt} gives $\mathcal{X}\phi_+^{opt} = -(\mathcal{V}_+^{-1}\mathcal{F}_+ - \mathcal{L})\phi_-^{opt}$. These define the optimal singular vector from (4.22) as

$$u_{opt}(z) = \frac{V_+(z)((F(z)-\mathcal{L})\phi_-^{opt} + \mathcal{X}\phi_+^{opt})}{\chi_{\gamma_{opt}}(z)}, \tag{4.33}$$

$$= \frac{V_+(z)(F(z)-\mathcal{V}_+^{-1}\mathcal{F}_+)\phi_-^{opt}}{\chi_{\gamma_{opt}}(z)}. \tag{4.34}$$

Then, this gives $s_{opt}(z) = w_1(z) - m(z)q_1^{opt}(z)$ from Theorem 17:

$$s_{opt}(z) = \frac{(b(\mathbf{T})u_{opt})(z)}{(k(\mathbf{T})u_{opt})(z)} = \frac{b(z)u_{opt}(z) - m(z)V_+(z)\mathcal{B}\phi_-^{opt}}{k(z)u_{opt}(z) - m(z)V_+(z)\mathcal{K}\phi_-^{opt}}. \tag{4.35}$$

Theorem 20 *Under the assumptions of Theorem 18 the optimal controller for the plant $p(z) = m(z)n_1(z)$ is given by*

$$c_{opt}(z) = \left(\frac{w_1(z)w_1(z^{-1})}{{\gamma_{opt}}^2} - 1\right)\frac{g_{opt}(z)}{1 + m(z)g_{opt}(z)}n_1^{-1}(z), \tag{4.36}$$

where

$$g_{opt}(z) := {\gamma_{opt}}^2\,\frac{\tilde{k}(z)}{b(z)}\,\frac{V_+(z)(b(z)\mathcal{K} - k(z)\mathcal{B})\phi_-^{opt}}{V_+(z)\mathcal{V}_+^{-1}\mathcal{F}_+\phi_-^{opt}}, \tag{4.37}$$

and non-zero ϕ_-^{opt} satisfies $\mathcal{R}_{\gamma_{opt}}\phi_-^{opt} = 0$.

Proof. Solve for q_1^{opt} from $q_1^{opt} = (w_1 - s_{opt})/m$, where s_{opt} is given by (4.35) in terms of u_{opt}, which is expressed in terms of ϕ_-^{opt} by (4.34). Therefore, we can obtain c_{opt} from (4.9)

$$c_{opt}(z) = \frac{q_{opt}(z)}{1 - p(z)q_{opt}(z)} = \frac{q_1^{opt}(z)n_1^{-1}(z)}{w_1(z) - m(z)q_1^{opt}(z)}$$

in terms of ϕ_-^{opt} by substitution. After algebraic manipulations this substitution gives us the simplified formulae (4.36-4.37). □

The structure of the controller given by (4.36) was first observed in [82] and the general form for $g_{opt}(z)$ was obtained in [75].

Theorems 19 and 20 provide an explicit solution to the one-block $\mathcal{H}^\infty$ control problem for stable distributed plants. The results of this section can be summarized as follows.

1. γ_{opt} is the norm of the Sarason operator $w_1(\mathbf{T})$.

2. The singular value/singular vector equation for this operator implies that there exists a singular vector $u \in \mathcal{H}(m)$ for a singular value candidate ρ if and only if

   $$\mathbf{A}_\rho u = 0.$$

3. The equation $\mathbf{A}_\rho u = 0$ is equivalent to

$$\chi_\rho(z)u(z) = R_\rho(z)\phi,$$

where ϕ is a constant in $\mathbb{C}^{2n}$, $\chi_\rho(z)$ is a polynomial of degree $2n$ and the entries of $1 \times 2n$ vector $R_\rho(z)$ can be explicitly computed.

4. From 3, we have $2n$ linear equations for the existence of a non-zero $u \in \mathcal{H}(m)$

$$R_\rho(z_i)\phi = 0, \quad \text{where } 0 \neq \phi \in \mathbb{C}^{2n},$$

and $\chi(z_i) = 0, \quad i = 1, 2, \dots 2n.$

5. The largest value of ρ for which there is such a non-zero ϕ gives γ_{opt}.

6. The optimal controller can be obtained from a non-zero ϕ_{opt} satisfying the above $2n$–equations for $\rho = \gamma_{opt}$. The solving of the above $2n$ equations is easily reduced to that of a system of n equations.

4.3.6 Example

This section contains an example which illustrates the numerical computation of γ_{opt} from the formula obtained in Theorem 19.

Let us consider the weighting function

$$W_1(s) = \frac{0.1(s + 0.5)(s + 100)}{(s + 1)(s + 5)},$$

or equivalently on the unit disc

$$w_1(z) = W_1\left(\frac{1 + z}{1 - z}\right) = \frac{b(z)}{k(z)} = \frac{15.15 - 9.8z - 4.95z^2}{12 - 8z}.$$

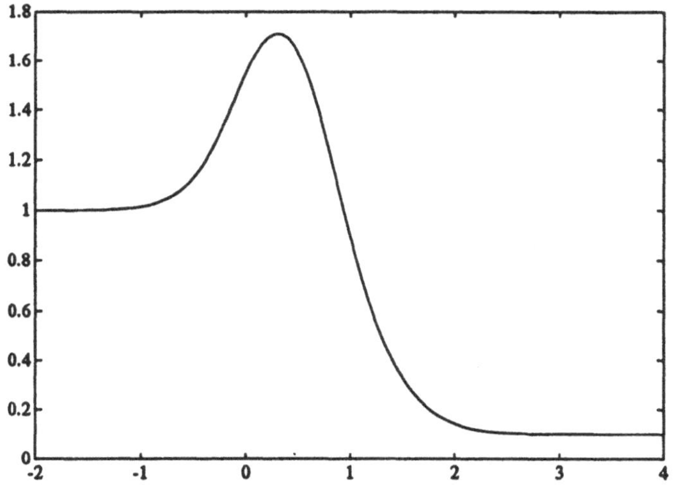

Figure 4.1: $|W_1(j\omega)|$ versus $\log(\omega)$

Suppose that the inner part of the plant is given by

$$M(s) = e^{-hs}\frac{a-s}{a+s}, \quad m(z) = M(\frac{1+z}{1-z}),$$

where h is the amount of time delay, and a is the non-minimum phase zero of the plant to be controlled. We will investigate how γ_{opt} changes as h and a vary.

The magnitude plot for the weight is given in Figure 4.1, which shows that the magnitude of the weight is relatively large in the low frequency range up to 20rad/sec.

When $h > 0$ the essential norm is $\|w_1(\mathbf{T})\|_e = w_1(1) = 0.1$, because $z = 1$ is the only essential singularity for $m(z) := M(\frac{1+z}{1-z})$. It is easy to see from (4.8) that $|w_1(\frac{a-1}{a+1})|$ (i.e. $|W_1(a)|$) is also a lower bound for γ_{opt}, since $m(\frac{a-1}{a+1}) = 0$ (i.e. $M(a) = 0$). An upper bound for γ_{opt} is $\|w_1\|_\infty = 1.71$, which can be seen from Figure 4.1. We have constructed the matrix $\mathcal{R}_\rho$ for the values of ρ between 0.1 (the essential norm) and 1.71. The minimum singular value of $\mathcal{R}_\rho$ is plotted, for several different

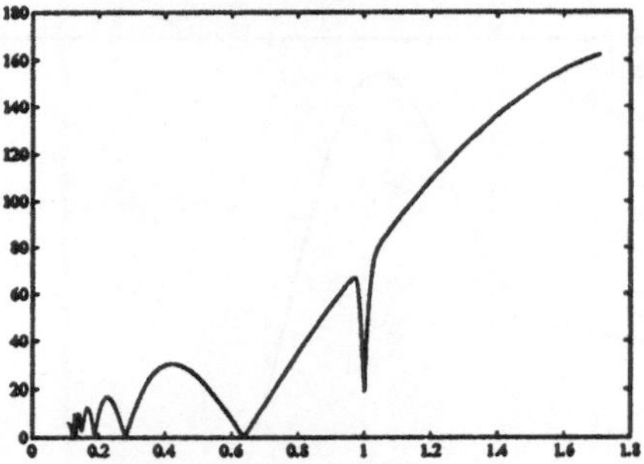

Figure 4.2: $\sigma_{min}(\mathcal{R}_\rho)$ versus ρ for $h = 0.1$, $a = 100$

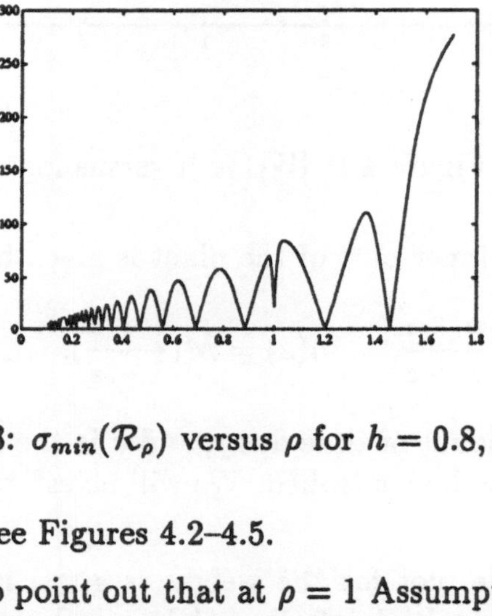

Figure 4.3: $\sigma_{min}(\mathcal{R}_\rho)$ versus ρ for $h = 0.8$, $a = 100$

values of h and a, see Figures 4.2–4.5.

We would like to point out that at $\rho = 1$ Assumption 4.1 is not satisfied because χ_ρ has double roots at $z = 1$. This is the reason why the plot $\sigma_{min}(\mathcal{R}_\rho)$ shows a zero at $\rho = 1$. In other words $\rho = 1$ is a degenerate point and should be discarded. The other points where $\sigma_{min}(\mathcal{R}_\rho)$ becomes zero indicates a singular value of $w_1(\mathbf{T})$ (also a singular value of the Hankel operator $\Gamma_{m^* w_1}$, by virtue of Theorem 16). Another interesting point to observe is that the singular values accumulate at the essential norm 0.1.

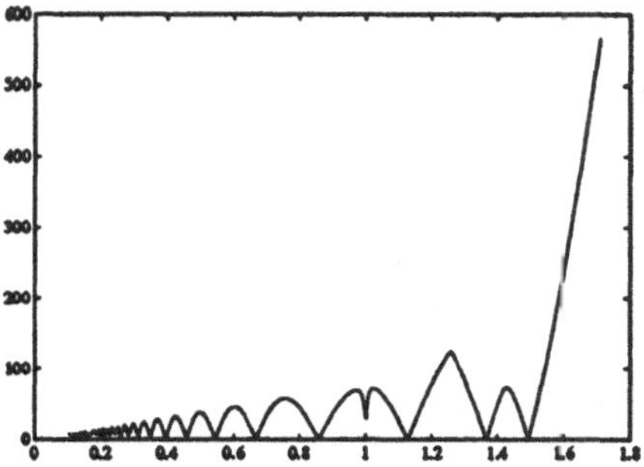

Figure 4.4: $\sigma_{min}(\mathcal{R}_\rho)$ versus ρ for $h = 0.8$, $a = 1$

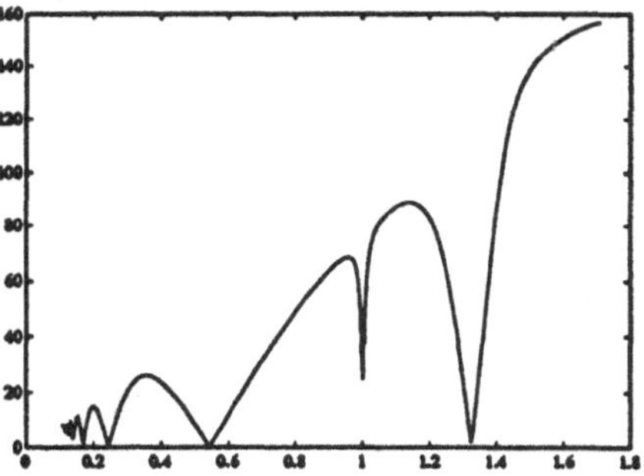

Figure 4.5: $\sigma_{min}(\mathcal{R}_\rho)$ versus ρ for $h = 0.1$, $a = 1$

(a,h):	(100 , 0.1)	(100 , 0.8)	(1 , 0.8)	(1 , 0.1)
γ_{opt}:	0.56	1.46	1.49	1.33
$\lvert W_1(a)\rvert$:	0.19	0.19	1.26	1.26

Table 4.1: γ_{opt} versus a and h

The plots shown in Figures 4.2–4.5 give us Table 4.1, which indicates γ_{opt} for different values of h and a.

We see that as h increases γ_{opt} increases for fixed a. The large values of a (e.g. $a = 100$) does not effect γ_{opt} significantly. Note that $|W_1(a)|$ is small when a is large, and in this case the relative effect of a on γ_{opt} is small. Conversely if a is small then $|W_1(a)|$ is large and in this case γ_{opt} is mainly determined by a.

It is well known that time delays and non-minimum phase zeros have a negative effect on the performance of the system. Here we can quantify this effect precisely relative to the magnitude of the weighting function.

4.3.7 Special Case: Finite Dimensional Plants

We now present a simple formula for the computation of γ_{opt} and the corresponding optimal interpolant q_1^{opt} for the finite dimensional case where the plant and the weight are rational. For this special case we consider the original problem data in terms of functions defined on $\mathbb{C}_+$. Suppose $R = M^*W_1$ is a rational function in $\mathcal{L}^\infty(j\mathrm{R})$, and consider the Hankel operator $\mathbf{\Gamma}_R$ defined from $\mathcal{H}^2$ to $\mathcal{L}^2 \ominus \mathcal{H}^2$ as

$$\mathbf{\Gamma}_R v = \mathbf{P}_- R v \quad \text{for} \quad v \in \mathcal{H}^2 ,$$

where $\mathbf{P}_-$ is the orthogonal projection from $\mathcal{L}^2$ to $\mathcal{L}^2 \ominus \mathcal{H}^2$. Then the Nehari theorem implies that

$$\gamma_{opt} = \inf_{Q_1 \in \mathcal{H}^\infty(\mathbb{C}_+)} \|R - Q_1\|_\infty = \|\mathbf{\Gamma}_R\| .$$

In the special case where

$$R(s) = \sum_{k=1}^{n} \frac{d_k}{a_k - s}, \quad \text{where} \quad \mathrm{Re}(a_k) > 0$$

we claim that

$$\gamma_{opt} = \|\mathbf{\Gamma}_R\| = \left(\lambda_{max}(A^* D^* A D)\right)^{\frac{1}{2}} \tag{4.38}$$

where $\lambda_{max}(\cdot)$ denotes the largest eigenvalue, and

$$A = \begin{bmatrix} \frac{1}{a_1+\bar{a}_1} & \cdots & \frac{1}{a_n+\bar{a}_1} \\ \vdots & & \vdots \\ \frac{1}{a_1+\bar{a}_n} & \cdots & \frac{1}{a_n+\bar{a}_n} \end{bmatrix}, \quad D = \begin{bmatrix} d_1 & 0 & 0 \\ 0 & \ddots & 0 \\ 0 & 0 & d_n \end{bmatrix}.$$

In order to prove this fact consider the singular value/singular vector equation

$$\sigma^2 g = \mathbf{\Gamma}_R^* \mathbf{\Gamma}_R g \tag{4.39}$$

for a non-zero singular vector $g \in \mathcal{H}^2$, corresponding to a singular value σ. The right hand side of the above equation can be expressed as follows. First note that

$$\mathbf{\Gamma}_R g = [\frac{d_1}{a_1 - s} \quad \cdots \quad \frac{d_n}{a_n - s}] x_g \tag{4.40}$$

where $x_g := [g(a_1) \ \cdots \ g(a_n)]^T$. Then, using the Cauchy formula

$$\langle f \,, \frac{1}{a - s} \rangle = f(a) \quad \text{for } f \in \mathcal{H}^2 \text{ and } \mathrm{Re}(a) > 0\,,$$

it is easy to show that (4.39) holds if and only if

$$\sigma^2 g(s) = [\frac{\bar{d}_1}{\bar{a}_1 + s} \quad \cdots \quad \frac{\bar{d}_n}{\bar{a}_n + s}] A D x_g\,. \tag{4.41}$$

Clearly (4.41) implies that $g \in \mathcal{H}^2$ is non-zero if and only if x_g is non-zero. But by evaluating (4.41) at a_k's we obtain n equations in n unknowns (components of x_g). This set is

$$A^*D^*ADx_g = \sigma^2 x_g \; .$$

Thus, the largest singular value of $\mathbf{\Gamma}_R$ is the square root of the largest eigenvalue of A^*D^*AD. Once $\gamma_{opt} = \|\mathbf{\Gamma}_R\|$ is obtained this way the optimal interpolant Q_1 is given by

$$R - Q_1 = \frac{\mathbf{\Gamma}_R g_{max}}{g_{max}} \tag{4.42}$$

where $g_{max}(s)$ is given by (4.41) with $\sigma = \gamma_{opt}$, and x_g is a non-zero eigenvector for A^*D^*AD corresponding to the eigenvalue γ_{opt}^2. The right hand side of (4.42) can be easily computed from (4.40) and (4.41).

Chapter 5

$\mathcal{H}^\infty$ Control of Unstable Plants

In Chapter 4, we have seen that the one block $\mathcal{H}^\infty$ optimal control problem can be solved by studying the singular values and vectors of a Hankel (or Sarason) operator. The main result of the previous chapter was that the existence of a singular vector for this Hankel operator is equivalent to the existence of a non-zero solution to finitely many linear equations given in Theorem 18. Moreover, the corresponding singular vector (and hence the optimal controller) can be constructed from these finitely many linear equations. In this chapter we will extend these results to the two block $\mathcal{H}^\infty$ optimal control problem for infinite dimensional unstable plants. We will derive finitely many linear equations equivalent to singular value/singular vector equation of a "two block operator" which gives the $\mathcal{H}^\infty$ controller in this case.

5.1 Two block operator

In this section, we will consider the two block problem defined in Chapter 3, and define the corresponding two block operator. As usual we assume that the problem data is transformed to unit disc via the conformal map $s = \frac{1+z}{1-z}$. So $p(z) = P(\frac{1+z}{1-z})$ represents the transfer function of the plant, and similarly for $w_1(z)$, $w_2(z)$ and $c(z)$.

Now recall the two-block $\mathcal{H}^\infty$ control problem defined in Chapter 3.

$$\gamma_{opt} = \inf_{[c,p]\ stable} \left\| \begin{bmatrix} w_1(1+pc)^{-1} \\ w_2pc(1+pc)^{-1} \end{bmatrix} \right\|_\infty , \tag{5.1}$$

where the weight w_1 comes from the sensitivity reduction condition, and $w_2 = w_m$ is the multiplicative uncertainty bound; see Section 3.6.1. We will use the controller parameterization of Section 3.2:

$$c = \frac{x+dq}{y-nq},$$

where $p = n/d$, with $n, d \in \mathcal{H}^\infty$, and $x, y \in \mathcal{H}^\infty$ satisfy

$$xn + yd = 1. \tag{5.2}$$

In Chapter 3, we assumed that $d(z) = m_d(z)$ is a rational inner function with $m_d(0) \neq 0$. In other words the denominator of p is a finite Blaschke product of the form

$$m_d(z) = \prod_{k=1}^{\ell} \left(\frac{z-a_k}{1-\overline{a_k}z} \right) \quad |a_k| < 1, \quad \forall\ k = 1, 2, \ldots, \ell.$$

This means that the unstable poles of the plant are finitely many and distinct. Recall also from Chapter 3 that $x(z)$ can be chosen as a rational function in (5.2).

With Assumptions 3.1 and 3.2 of Chapter 3, and using the above controller parameterization, it is easy to show that the two-block problem (5.1) is equivalent to

$$\begin{aligned} \gamma_{opt} &= \inf_{q \in \mathcal{H}^\infty(\mathsf{D})} \left\| \begin{bmatrix} w_1 \\ 0 \end{bmatrix} - \begin{bmatrix} w_1 n_2 \\ -w_2 n_2 \end{bmatrix} m_n n_1 (x + m_d q) \right\|_\infty \\ &= \inf_{q_1 \in \mathcal{H}^\infty(\mathsf{D})} \left\| \begin{bmatrix} w_1 \\ 0 \end{bmatrix} - \begin{bmatrix} w_1 n_2 \\ -w_2 n_2 \end{bmatrix} m_n (x n_1 + m_d q_1) \right\|_\infty , \end{aligned}$$

where $q_1 = n_1 q$ or $q = q_1/n_1$. Note that $q \in \mathcal{H}^\infty(\mathsf{D})$ if and only if $q_1 \in \mathcal{H}^\infty(\mathsf{D})$.

It is easy to choose a polynomial $r(z)$ such that that

$$g_1(z) := \frac{r(z) - x(z)n_1(z)}{m_d(z)} \in \mathcal{H}^\infty(\mathsf{D}).$$

Then, defining $q_2 = q_1 - g_1$ (there is an invertible relationship between q_1 and q_2, hence between q and q_2) we have

$$\gamma_{opt} = \inf_{q_2 \in \mathcal{H}^\infty(\mathsf{D})} \left\| \begin{bmatrix} w_1 \\ 0 \end{bmatrix} - \begin{bmatrix} w_1 n_2 \\ -w_2 n_2 \end{bmatrix} m_n(r + m_d q_2) \right\|_\infty . \tag{5.3}$$

Note that the only infinite dimensional part in (5.3) is the inner function m_n. Also recall from Assumption 3.2 that $w_3 := w_2 n_2$ is in $\mathcal{H}^\infty(\mathsf{D})$ and so is w_3^{-1}. The problem (5.3) can further be reduced as follows. First from the assumptions of Section 3.6.1 it follows that $w_1 n_2$ and n_3 are rational functions therefore we can perform a spectral factorization

$$n_2^* w_1^* w_1 n_2 + w_3^* w_3 =: g^* g \tag{5.4}$$

with a rational function $g(z)$ such that $g, g^{-1} \in \mathcal{H}^\infty(\mathsf{D})$. Such function g exists because $w_3^{-1} \in \mathcal{H}^\infty(\mathsf{D})$, and g can be obtained from w_1, n_2 and w_2, see e.g. [4] and [39].

Note that the matrix

$$L = \begin{bmatrix} \frac{w_1 n_2}{g} & \frac{w_3^*}{g^*} \\ \frac{-w_3}{g} & \frac{n_2^* w_1^*}{g^*} \end{bmatrix} \tag{5.5}$$

is unitary. Hence from (5.3) we have

$$\gamma_{opt} = \inf_{q_2 \in \mathcal{H}^\infty(\mathsf{D})} \left\| L^* \left(\begin{bmatrix} w_1 \\ 0 \end{bmatrix} - \begin{bmatrix} w_1 n_2 \\ -w_2 n_2 \end{bmatrix} m_n(r + m_d q_2) \right) \right\|_\infty$$

which leads to

$$\gamma_{opt} = \inf_{q_2 \in \mathcal{H}^\infty(\mathsf{D})} \left\| \begin{bmatrix} \frac{w_1 w_1^* n_2^*}{g^*} \\ -\frac{\bar{w}_1 w_3}{g} \end{bmatrix} - \begin{bmatrix} g m_n \\ 0 \end{bmatrix} (r + m_d q_2) \right\|_\infty . \tag{5.6}$$

Note that the second block in (5.6) is

$$g_0 := -w_1 w_3 / g,$$

and it is independent of the free parameter q_2. Also, since $w_3, g^{-1} \in \mathcal{H}^\infty(\mathsf{D})$, we have that $g_0 \in \mathcal{H}^\infty(\mathsf{D})$. In order to express (5.6) in terms of $\mathcal{H}^\infty(\mathsf{D})$ functions we find a rational inner function $m_w(z)$ such that the rational function

$$w_0 := m_w \frac{w_1 w_1^* n_2^*}{g^*} \quad \text{is in} \quad \mathcal{H}^\infty(\mathsf{D}).$$

Since w_1, n_2, g are rational we can always find such $m_w(z)$. In fact one can construct $m_w(z)$ by taking its zeros as $1/\overline{p_i}$ and $1/\overline{z_i}$ where p_i's are the poles of $w_1 n_2$, and z_i's are the zeros of g. Note that these poles and zeros are outside $\overline{\mathsf{D}}$. Then, from (5.6), we have

$$\gamma_{opt} = \inf_{q_2 \in \mathcal{H}^\infty(\mathsf{D})} \left\| \begin{bmatrix} m_w & 0 \\ 0 & 1 \end{bmatrix} \left(\begin{bmatrix} \frac{w_1 w_1^* n_1^*}{g^*} \\ -\frac{w_1 w_3}{g} \end{bmatrix} - \begin{bmatrix} g m_n \\ 0 \end{bmatrix} (r + m_d q_2) \right) \right\|_\infty ,$$

which can be re-written as

$$\gamma_{opt} = \inf_{q_3 \in \mathcal{H}^\infty(\mathsf{D})} \left\| \begin{bmatrix} w_0 - m_1 \widehat{w}_0 - m_1 m_2 q_3 \\ g_0 \end{bmatrix} \right\|_\infty , \tag{5.7}$$

where

$$\begin{aligned} g_0 &:= -w_1 w_3 / g, \\ w_0 &:= m_w w_1 w_1^* n_2^* / g^*, \\ \widehat{w}_0 &:= g r, \\ m_1 &:= m_n m_w \\ m_2 &:= m_d \\ q_3 &:= g q_2 \quad \text{or} \quad q_2 = q_3 / g, \end{aligned}$$

with $w_3 := (w_2 n_2)$. Note once more that $w_0, g_0, \widehat{w}_0, m_2$ are rational functions in $\mathcal{H}^\infty(\mathsf{D})$, with m_2 inner, and m_1 is arbitrary inner. Moreover, when the plant is stable we can choose $m_d = 1$, $y = 1$, $x = 0$ in (5.2) and in this case we have $\widehat{w}_0 = 0$.

Let m denote the inner function $m_1 m_2 = m_w m_n m_d$. and define the "two block operator" $\mathbf{A} : \mathcal{H}^2(\mathbf{D}) \to \mathcal{H}(m) \oplus \mathcal{H}^2(\mathbf{D})$

$$\mathbf{A} := \begin{bmatrix} \mathbf{P}_{\mathcal{H}(m)}(w_0(\mathbf{S}) - m_1(\mathbf{S})\widehat{w}_0(\mathbf{S})) \\ g_0(\mathbf{S}) \end{bmatrix}.$$

Then we have the following result which can be seen as the two block version of Theorem 16.

Theorem 21 *Notation as above, we have*

$$\gamma_{opt} = \|\mathbf{A}\|. \tag{5.8}$$

Proof. For notational convenience we introduce $w := w_0 - m_1\widehat{w}_0$. Observe that for any $q_3 \in \mathcal{H}^\infty(\mathbf{D})$ we have

$$\begin{aligned} \left\| \begin{bmatrix} w - mq_3 \\ g_0 \end{bmatrix} \right\|_\infty &= \left\| \begin{bmatrix} \mathbf{P}_+(w(\mathbf{S}) - m(\mathbf{S})q_3(\mathbf{S})) \\ g_0(\mathbf{S}) \end{bmatrix} \right\| \\ &\geq \left\| \begin{bmatrix} \mathbf{P}_{\mathcal{H}(m)}(w(\mathbf{S}) - m(\mathbf{S})q_3(\mathbf{S})) \\ g_0(\mathbf{S}) \end{bmatrix} \right\| \\ &= \left\| \begin{bmatrix} \mathbf{P}_{\mathcal{H}(m)}w(\mathbf{S}) \\ g_0(\mathbf{S}) \end{bmatrix} \right\| = \|\mathbf{A}\|. \end{aligned}$$

Therefore, $\gamma_{opt} \geq \|\mathbf{A}\|$. In order to complete the proof we need to show the existence of $q_3^{opt} \in \mathcal{H}^\infty(\mathbf{D})$ such that

$$\left\| \begin{bmatrix} w - mq_3^{opt} \\ g_0 \end{bmatrix} \right\|_\infty = \|\mathbf{A}\|.$$

The key observation for this step is

$$\mathbf{AS} = \begin{bmatrix} \mathbf{TP}_{\mathcal{H}(m)}w(\mathbf{S}) \\ \mathbf{S}g_0(\mathbf{S}) \end{bmatrix},$$

where $\mathbf{T}$ is the compression of $\mathbf{S}$ on $\mathcal{H}(m)$. Then the result follows from the commutant lifting theorem [94], [28]. The details can be found in Chapter 8 even for the more general case; see also [34]. □

With Theorem 21, the two-block problem is reduced to a norm computation problem. The optimal performance level γ_{opt} is the norm of $\mathbf{A}$. The essential norm of this operator was computed in [33, 79, 81, 120] as

$$\|\mathbf{A}\|_e = \max\left\{\|g_0\|_\infty \ , \ \left\| \begin{bmatrix} w_0(z_i) \\ g_0(z_i) \end{bmatrix} \right\|\right\} \tag{5.9}$$

where z_i runs over the essential singularities of m_n. Assuming the norm is achieved at a singular value (i.e. $\gamma_{opt} > \|\mathbf{A}\|_e$) the optimal controller can be computed from a non-zero singular vector $x_o \in \mathcal{H}^2(\mathsf{D})$ satisfying

$$\mathbf{A}^*\mathbf{A}x_o = \gamma_{opt}^2 x_o. \tag{5.10}$$

Therefore, we need to understand the "action of" $\mathbf{A}^*\mathbf{A}$ on an element of $\mathcal{H}^2(\mathsf{D})$, in order to derive necessary and sufficient conditions for the existence of a candidate singular vector x_o associated with the maximum singular value γ_{opt} of $\mathbf{A}$.

It is easy to verify that $\mathbf{A}^*\mathbf{A}$ can be expressed in terms of Hankel and Toeplitz operators:

$$\mathbf{A}^*\mathbf{A} = \mathbf{\Gamma}^*_{m^*w}\mathbf{\Gamma}_{m^*w} + \mathbf{\Upsilon}^*_{g_0}\mathbf{\Upsilon}_{g_0}.$$

This type of "Hankel + Toeplitz" operators for several different special cases of m, w and g_0 have been studied to obtain solutions to two block $\mathcal{H}^\infty$ control problems for several different special types of plants, see for example, [27] [108] [120], etc.

5.2 Reduction to one block

We can solve the two block problem (5.1) by reducing (5.10) to an equation of the type (4.11), which corresponds to a one block problem. This technique has been used widely in the early $\mathcal{H}^\infty$ control literature, see e.g., [39] as well as [108] where the two block problem is reduced a broadband matching problem of [55]. First step in this procedure is to write the singular value/singular vector equation: ρ^2 is an eigenvalue, with finite multiplicity, of $\mathbf{A}^*\mathbf{A}$ if and only if there exists a non-zero $x \in \mathcal{H}^2(\mathrm{D})$ such that

$$\left(\rho^2\mathbf{I} - \mathbf{A}^*\mathbf{A}\right) x = 0$$

which is equivalent to

$$\left(\rho^2\mathbf{I} - g_0(\mathbf{S})^* g_0(\mathbf{S})\right) x = w(\mathbf{S})^* \mathbf{P}_{\mathcal{H}(m)} w(\mathbf{S}) x. \tag{5.11}$$

It is clear from (5.9) that $\|g_0\|_\infty$ is a lower bound for γ_{opt}. Since we are interested in the *largest* singular value for $\mathbf{A}$, we will assume that $\rho > \|g_0\|_\infty$. Then, there exists a *rational* function (which depends on ρ) $f_\rho \in \mathcal{H}^\infty(\mathrm{D})$ with $f_\rho^{-1} \in \mathcal{H}^\infty(\mathrm{D})$ such that

$$f_\rho(\mathbf{S})^* f_\rho(\mathbf{S}) = \rho^2\mathbf{I} - g_0(\mathbf{S})^* g_0(\mathbf{S}).$$

In fact f_ρ can be computed from rational spectral factorization techniques, see e.g. [39]. Now defining $y = f_\rho(\mathbf{S})x$, i.e. $x = f_\rho(\mathbf{S})^{-1}y$, the equation (5.11) becomes

$$y = w_\rho(\mathbf{S})^* \mathbf{P}_{\mathcal{H}(m)} w_\rho(\mathbf{S}) y, \tag{5.12}$$

where $w_\rho := f_\rho^{-1} w = w_{0,\rho} - m_1 \widehat{w}_{0,\rho}$, with $w_{0,\rho} := f_\rho^{-1} w_0$ and $\widehat{w}_{0,\rho} := f_\rho^{-1} \widehat{w}_0$. Let $\mathbf{T}$ be the compression of the shift operator, defined on $\mathcal{H}(m)$. Since $\mathbf{T}^* = \mathbf{S}^*$ on $\mathcal{H}(m)$, the right hand side of (5.12) is in $\mathcal{H}(m)$, this implies that y has to be in $\mathcal{H}(m)$, and hence (5.12) is equivalent to

$$(1 \cdot \mathbf{I} - w_\rho(\mathbf{T})^* w_\rho(\mathbf{T}))\, y = 0. \tag{5.13}$$

In other words 1 has to be a singular value of the Sarason's operator $w_\rho(\mathbf{T})$. The equation (5.13) corresponds to a one block problem. In the following sections we discuss the solution of this problem.

5.3 Stable plant case

Recall that when the plant is stable we can choose $r(z) = 0$ and consequently we have $\widehat{w}_0 = 0$. Therefore, in this case $w_\rho = w_{0,\rho}$ which is *rational.* Hence there exist polynomials $b_\rho(z)$ and $k_\rho(z)$ with $k_\rho^{-1}(z) \in \mathcal{H}^\infty(\mathbf{D})$ such that $w_\rho = b_\rho(z)/k_\rho(z)$. By defining $y' = k_\rho^{-1}(\mathbf{T})y$ we see that (5.13) is equivalent to

$$(k_\rho(\mathbf{T})^* k_\rho(\mathbf{T}) - b_\rho(\mathbf{T})^* b_\rho(\mathbf{T}))\, y' = 0. \tag{5.14}$$

Note that the left hand side of (5.14) is a skew Toeplitz operator acting on an element y' of $\mathcal{H}(m)$. In Chapter 4 we have solved the problem (4.12), which is the same as (5.14). Thus, the results of Chapter 4 directly apply to two block $\mathcal{H}^\infty$ optimal control problem for stable plants.

5.4 Unstable plant case

In the general case where the plant satisfies Assumption 3.1 we have $\widehat{w}_{0,\rho} \neq 0$, and hence w_ρ is not a rational function, so we cannot directly apply the results of Chapter 4. Nevertheless, w_ρ is not arbitrary, it has a special structure:

$$w_\rho = w_{0,\rho} - m_1 \widehat{w}_{0,\rho}$$

with $w_{0,\rho}$ and $\widehat{w}_{0,\rho}$ *rational* functions in $\mathcal{H}^\infty(\mathbf{D})$, and m_1 related to m by: $m = m_1 m_2$, where m_2 is a *finite* Blaschke product. So, there exist *polynomials* $b_\rho(z)$, $c_\rho(z)$ and $k_\rho(z)$ such that $w_{0,\rho} = b_\rho/k_\rho$ and $\widehat{w}_{0,\rho} = c_\rho/k_\rho$. In this case defining $y' = k_\rho^{-1}(\mathbf{T})y$ we see that (5.13) is equivalent to

$$\begin{aligned}\Big(k_\rho(\mathbf{T})^* k_\rho(\mathbf{T}) - b_\rho(\mathbf{T})^* b_\rho(\mathbf{T})\Big) y' = \Big(&- b_\rho(\mathbf{T})^* c_\rho(\mathbf{T}) m_1(\mathbf{T}) \\ &- m_1(\mathbf{T})^* c_\rho(\mathbf{T})^* (b_\rho(\mathbf{T}) - m_1(\mathbf{T}) c_\rho(\mathbf{T}))\Big) y'. \end{aligned} \tag{5.15}$$

The left hand side of (5.15) is the same as the left hand side of the equation (5.14). It is in the form of a skew Toeplitz operator acting on an element of $\mathcal{H}(m)$. From Chapter 4 we know exactly how this operator acts on y'. As for the right hand side of (5.15), again similarly to the formulae given in Chapter 4 we can explicitly write down the action of $b_\rho(\mathbf{T})$, $b_\rho(\mathbf{T})^*$, $c_\rho(\mathbf{T})$ and $b_\rho(\mathbf{T})^*$ on an element of $\mathcal{H}(m)$. We need to derive a similar formulae for the action of $m_1(\mathbf{T})$ and $m_1(\mathbf{T})^*$ on elements of $\mathcal{H}(m)$.

Note that y' is in $\mathcal{H}(m) = \mathcal{H}(m_1) \oplus m_1\mathcal{H}(m_2)$. Hence y' has an orthogonal decomposition of the form $y' = u + m_1 v$ where $u \in \mathcal{H}(m_1)$ and $v \in \mathcal{H}(m_2)$. Recall from Chapter 2 that $f_1, \ldots, f_\ell$ form a basis for $\mathcal{H}(m_2)$, where $f_i(z) = (1 - \overline{a_i} z)^{-1}$. Therefore, v has to be in the form

$$v(z) = [f_1(z) \quad \ldots \quad f_\ell(z)] \begin{bmatrix} \alpha_1 \\ \vdots \\ \alpha_\ell \end{bmatrix} \tag{5.16}$$

for some constants $\alpha_1, \ldots, \alpha_\ell$. Now we can compute $m_1(\mathbf{T})^* y'$ explicitly as follows:

$$m_1(\mathbf{T})^* y' = \mathbf{P}_+ m_1^* y' = \mathbf{P}_+(m_1^* u + v) = v \tag{5.17}$$

where we have used the fact that $m_1^* u \in \mathcal{L}^2 \ominus \mathcal{H}^2$, since $u \in \mathcal{H}(m_1)$. Similarly, we can compute $m_1(\mathbf{T}) y'$ in terms of finitely many constants as follows. Note that

$$m_1(\mathbf{T}) y' = m_1 y' - m_1 m_2 \mathbf{P}_+ m_2^* y' = m_1 \mathbf{P}_{\mathcal{H}(m_2)} y'.$$

Since $\mathcal{H}(m_2)$ is finite dimensional, from Lemma 4 we have

$$(m_1(\mathbf{T}) y')(z) = m_1(z)\ [f_1(z) \quad \ldots \quad f_\ell(z)] \Lambda^{-1} \begin{bmatrix} \beta_1 \\ \vdots \\ \beta_\ell \end{bmatrix} \tag{5.18}$$

where Λ is as defined in Chapter 2, and $\beta_i := y'(a_i)$ is a constant, $i = 1, \ldots, \ell$.

With the observations (5.17) and (5.18), we can write down the necessary and sufficient conditions for a non-zero $y' \in \mathcal{H}(m)$ to satisfy (5.15). Assuming that the maximum of the degrees of the polynomials b_ρ, c_ρ and k_ρ is n (and n is the degree of k_ρ), the unknown constants which determine the existence of a singular vector y' are $\phi_{-n},\ldots,\phi_{-1},\phi_0,\ldots,\phi_{n-1}$ and $\alpha_1,\ldots,\alpha_\ell$, $\beta_1,\ldots,\beta_\ell$, where ϕ_{-i}'s and ϕ_i's come from the power series expansions $(m^*y')(z) = \sum_{k=1}^{\infty}\phi_{-k}z^{-k}$ and $y'(z) = \sum_{k=0}^{\infty}\phi_k z^k$. Note that in case $n = 0$ there are no ϕ_k's to be determined. Accordingly, we define

$$\Phi := [\phi_{-n}^* \quad \cdots \quad \phi_{n-1}^* \;\; \alpha_1^* \quad \cdots \quad \alpha_\ell^* \;\; \beta_1^* \quad \cdots \quad \beta_\ell^*]^*$$

as the $2(n+\ell)\times 1$ vector of unknown constants. Recall that in the stable case there are $2n$ unknowns to be determined from $2n$ linear equations. In the unstable case the number of unknown constants increases by 2ℓ where ℓ is the number of unstable modes of the plant to be controlled.

It is easy to see from equations (4.22–4.24), and (5.17), (5.18) that the equation (5.15) can be rewritten as

$$\chi_\rho(z)y'(z) = R'_\rho(z)\Phi \tag{5.19}$$

where $\chi_\rho(z) = z^n(k_\rho(z^{-1})k_\rho(z)-b_\rho(z^{-1})b_\rho(z))$, and $R'_\rho(z)$ is a $1\times 2(n+\ell)$ vector valued function whose entries can be explicitly computed in terms of the problem data b_ρ, c_ρ, k_ρ, m_1, m_2. Similar to the stable case $2n$ equations are obtained by evaluating $R'_\rho(z)\Phi$ at the roots of the polynomial $\chi_\rho(z)$, whose order is $2n$. Extra ℓ equations are obtained by evaluating (5.19) at $a_1,\ldots,a_\ell$, and using the definition that $y'(a_i) = \beta_i$. The last set of ℓ equations are obtained by taking the orthogonal projection of both sides of (5.19) on $m_1\mathcal{H}^2$, noting that $m_1\mathcal{H}^2$ part of y' is defined as $m_1 v$, where v is defined in terms of α_i's as in (5.16). It was shown (cf. [79]) that these $2(n+\ell)$ equations determine a set of necessary and sufficient conditions for $y = k_\rho(\mathbf{T})y' \in \mathcal{H}(m)$ to be a singular vector of the Sarason operator $w_\rho(\mathbf{T})$ that satisfies (5.13). Again similar to the stable case, once we find a non-zero Φ satisfying these $2(n+\ell)$ linear equations we can construct a singular vector y, and from y we can obtain the optimal controller. The details of the derivation of these equations and the construction of the optimal controller can be found in [79] and [80]. For the convenience of the reader we present these $2(n+\ell)$ equations in a separate section at the end of this chapter. But first we study an example which demonstrates the main difference between the stable and unstable plant cases.

5.5 Example

In this section, we give a simple example to illustrate the previously described theory. We apply all the above computations to an unweighted mixed sensitivity minimization problem. In order to elucidate our methods, we will explicitly work through the required computations step by step.

Consider a plant $P(z) = m_n(z)/m_d(z)$, where m_n is arbitrary inner (possibly infinite dimensional) and m_d is a first order Blaschke function:

$$m_d(z) = \frac{z-a}{1-az}$$

with $a \in \mathbf{D}$ real and $m(a)$ real. The Bezout identity for this system is

$$Xm_n + Ym_d = 1,$$

so we can choose $X(z) = 1/m_n(a)$, constant. Using the parametrization $C = (X + m_dQ)/(Y - m_nQ)$, see Section 3.2, we can express the sensitivity $S = (1+PC)^{-1}$ and the complementary sensitivity $T = 1 - S$ functions in terms of the free parameter $Q \in \mathcal{H}^\infty$

$$S(z) = 1 - m_n(z)/m_n(a) - m_n(z)m_d(z)Q(z), \tag{5.20}$$

$$T = 1 - S(z) = m_n(z)/m_n(a) + m_n(z)m_d(z)Q(z). \tag{5.21}$$

In the unweighted mixed sensitivity, S and T, minimization problem (that is the two block problem defined in Section 3.6.1 with $W_1 = 1$ and $W_2 = 1$) we want to find

$$\gamma_{opt} = \inf_{Q\in\mathcal{H}^\infty} \left\| \begin{bmatrix} 1 - m_n(z)/m_n(a) \\ m_n(z)/m_n(a) \end{bmatrix} - \begin{bmatrix} 1 \\ -1 \end{bmatrix} m(z)Q \right\|_\infty \tag{5.22}$$

where $m(z) = m_n(z)m_d(z)$. By employing the inner/outer factorization for the constant matrix

$$\begin{bmatrix} 1 \\ -1 \end{bmatrix} = \begin{bmatrix} \frac{1}{\sqrt{2}} \\ -\frac{1}{\sqrt{2}} \end{bmatrix} \frac{1}{\sqrt{2}}$$

the above can be reduced to

$$\gamma_{opt} = \inf_{Q \in \mathcal{H}^\infty} \left\| \frac{1}{\sqrt{2}} \begin{bmatrix} 1 - 2m_n/m_n(a) - mQ \\ 1 \end{bmatrix} \right\|_\infty .$$

Therefore

$$\gamma_{opt} = \sqrt{\frac{1+\gamma_1^2}{2}},$$

where

$$\gamma_1 = \inf_{\tilde{Q} \in \mathcal{H}^\infty} \left\| 1 - 2m_n/m_n(a) - m\tilde{Q} \right\|_\infty . \qquad (5.23)$$

So the problem is reduced to computing γ_1, and the corresponding optimal interpolant. A lower bound for γ_1 can be computed by putting $z = a$ in the above equation, and an upper bound can be computed by choosing , say, $\tilde{Q} = 0$, i.e.

$$1 \le \gamma_1 \le \|1 - 2m_n(z)/m_n(a)\|_\infty .$$

By Sarason's theorem we have that $\gamma_1 = \|1 - 2m_n(\mathbf{T})/m_n(a)\|$, with $\mathbf{T} = \mathbf{P}_{\mathcal{H}(m)}\mathbf{S}|_{\mathcal{H}(m)}$. In order to compute the norm we form the singular value/singular vector equation

$$\left(\rho^2\mathbf{I} - (\mathbf{I} - 2m_n(\mathbf{T})^*/m_n(a))(\mathbf{I} - 2m_n(\mathbf{T})/m_n(a))\right) y = 0 \qquad (5.24)$$

where ρ^2 is a singular value with corresponding singular vector $y \in \mathcal{H}(m)$. Now we decompose y as $y = u + m_n v$, where $u \in \mathcal{H}(m_n)$, and $v \in \mathcal{H}(m_d)$.

We know the action of $m_n(\mathbf{T})^*$ and $m_n(\mathbf{T})$ on y:

$$m_n(\mathbf{T})^* y = v(z), \qquad m_n(\mathbf{T})y = m_n(z)\mathbf{P}_{\mathcal{H}(m_d)}y.$$

We can now write the equation (5.24) as follows. First note that $\mathcal{H}(m_d)$ is one dimensional and has a basis $f(z) = \frac{1}{1-az}$, so $v(z) = \alpha f(z)$ for some constant α, and moreover

$$\mathbf{P}_{\mathcal{H}(m_d)}y = \beta(1-a^2)f(z)$$

where $\beta := y(a)$ is a constant. We then have that (5.24) is equivalent to

$$(\rho^2-1)y = 4\beta\frac{1-a^2}{m_n(a)^2}f(z) - 2\beta\frac{1-a^2}{m_n(a)}m_n(z)f(z) - \frac{2\alpha}{m_n(a)}f(z). \tag{5.25}$$

Note that in this case we have $n = 0$ and $\ell = 1$. Hence the number of linearly independent equations that we obtain is $2(n+\ell) = 2$. Evaluating (5.25) at $z = a$ we obtain one of the equations as

$$(\rho^2-1)\beta = \frac{4\beta}{m_n(a)^2} - 2\beta - \frac{2\alpha}{m_n(a)}\frac{1}{1-a^2}. \tag{5.26}$$

The other equation is obtained by taking the orthogonal projection of (5.25) onto $m_n\mathcal{H}(m_d)$. After simplifications this can be found to be equivalent to

$$2\beta\frac{1-a^2}{m_n(a)} = (\rho^2+1)\alpha. \tag{5.27}$$

Then γ_1 is the largest value of $\rho \in [1\ ,\ \|1-2m_n(z)/m_n(a)\|_\infty]$ satisfying (5.26) and (5.27) for some non-zero constants α and β. This can easily be computed from (5.26) and (5.27), and the final answer is

$$\gamma_1^2 = \left(\frac{2}{m_n(a)^2} - 1\right) + \frac{2}{m_n(a)^2}\sqrt{1-m_n(a)^2}.$$

Consequently, for this example the optimal mixed sensitivity performance level $\gamma_{opt} = \sqrt{(1+\gamma_1^2)/2}$ can be computed as

$$\gamma_{opt} = \frac{1}{|m_n(a)|}\sqrt{1+\sqrt{1-m_n(a)^2}}.$$

The corresponding singular vector y can be found by computing a non-zero α and β satisfying (5.26) and (5.27). Then the optimal controller can be obtained from the vector y by using a method similar to that of Sarason given in Section 2.9.1 and by using the parametrization of all stabilizing controllers. More precisely, from the above we can compute that

$$y(z) = \left(\frac{1}{\gamma_1^2-1}\right)\left(\frac{2}{m_n(a)}\gamma_1^2 - (\gamma_1^2+1)m_n(z)\right)\alpha f(z).$$

Then $\tilde{Q}_{opt}$ satisfying

$$\begin{aligned}\gamma_1 &= \inf_{Q\in\mathcal{H}^\infty}\|1-2m_n(z)/m_n(a) - m_n(z)m_d(z)Q\|_\infty \\ &= \|1-2m_n(z)/m_n(a) - m_n(z)m_d(z)\tilde{Q}_{opt}(z)\|_\infty\end{aligned}$$

is given by the formula

$$((\mathbf{I} - \frac{2}{m_n(a)}m_n(\mathbf{T}))y)(z) = (1 - \frac{2m_n(z)}{m_n(a)} - m_n(z)m_d(z)\tilde{Q}_{opt}(z))y(z).$$

Using $m_n(\mathbf{T})y = m_n(z)\mathbf{P}_{\mathcal{H}(m_d)}y = \beta(1-a^2)f(z)m_n(z)$, we can solve for $\tilde{Q}_{opt}$:

$$\tilde{Q}_{opt}(z) = \frac{(\gamma_1^2-1)(\gamma_1^2+1) - 4\gamma_1^2/m_n(a)^2 + 2(\gamma_1^2+1)m_n(z)/m_n(a)}{m_d(z)(2\gamma_1^2/m_n(a) - (\gamma_1^2+1)m_n(z))}.$$

We are interested in finding Q_{opt} of (5.22) which is related to $\tilde{Q}_{opt}$ of (5.23) by $Q_{opt} = \frac{1}{2}\tilde{Q}_{opt}$. Employing the above formulae and (5.20) with $Q = Q_{opt}$, it is easy to compute that the optimal sensitivity is

$$S_{opt}(z) = \frac{1 - m_n(z)m_n(a)(\gamma_{opt}^4/\gamma_1^2)}{1 - m_n(z)m_n(a)(\gamma_{opt}^2/\gamma_1^2)}.$$

Hence the optimal controller is given by

$$\begin{aligned} C_{opt}(z) &= (\frac{1}{S_{opt}} - 1)\frac{m_d(z)}{m_n(z)} \\ &= m_d(z)\frac{m_n(a)(\gamma_{opt}^2 - 1)\gamma_{opt}^2/\gamma_1^2}{1 - m_n(z)m_n(a)\gamma_{opt}^4/\gamma_1^2}. \end{aligned}$$

One can check that at $z = a$ we have

$$1 - m_n(a)^2\frac{\gamma_{opt}^4}{\gamma_1^2} = 0$$

so that we do not have an unstable pole-zero cancellation in the product $P(z)C(z)$.

An important particular case of the above example is a plant (in continuous time) with a delay and one unstable pole:

$$P(s) = e^{-hs}\frac{\sigma s + 1}{\sigma s - 1}.$$

After transforming the data to the unit disc with the conformal map $z = \frac{s-1}{s+1}$, we find that

$$m_n(z) = e^{h\frac{z+1}{z-1}}, \quad m_d(z) = \frac{z - a}{1 - az},$$

with $a = (1 - \sigma)/(1 + \sigma)$. Then $m_n(a) = e^{-h/\sigma}$ and hence

$$\gamma_{opt} = e^{h/\sigma}\sqrt{1 + \sqrt{1 - e^{-2h/\sigma}}}.$$

It is interesting to note that as $h \to \infty$, and/or $\sigma \to 0$, the best achievable γ_{opt} increases exponentially, as expected.

Remark: The problem considered in this example is not very interesting from the practical control point of view. Indeed, the plant is not strictly proper (most realistic plants are), and usually there are non-constant weights associated with the sensitivity and complementary sensitivity. Our purpose in presenting this example was simply to elucidate our techniques, and to give some insights about the derivation of the $2(n+\ell)$ equations to compute the optimal performance and the controller. A more realistic example, involving an unstable delay system connected with the flight control of the X-29, has been worked-out in [20].

5.6 Explicit formulae for $2(n+\ell)$ equations

In this section, we summarize the $2(n+\ell)$ equations from [79]. Let us first recall the derivation of the two block problem data:

Plant: $p(z) = n_1(z)n_2(z)m_n(z)/m_d(z)$ where $n_1, n_1^{-1} \in \mathcal{H}^\infty$, m_n is inner, n_2 is rational outer, m_d is rational inner.

Bezout Identity: Using Lagrange interpolation find a rational function $x \in \mathcal{H}^\infty$ such that

$$y = \frac{1 - n_1 n_2 m_n x}{m_d} \in \mathcal{H}^\infty .$$

Weights: w_1 and w_2 are rational and such that $w_1, w_1^{-1} \in \mathcal{H}^\infty$, $w_3 := w_2 n_2 \in \mathcal{H}^\infty$, and $w_3^{-1} \in \mathcal{H}^\infty$.

Spectral Factorizations: Given n_2, w_1, w_2 and ρ, find g and f_ρ such that $g, g^{-1}, f_\rho, f_\rho^{-1} \in \mathcal{H}^\infty$ and

$$\begin{aligned} g^* g &= w_1^* n_2^* n_2 w_1 + w_3^* w_3 , \\ f_\rho^* f_\rho &= \rho^2 - \frac{w_1^* w_2^* w_2 w_1}{g^* g} . \end{aligned}$$

Interpolating Function: Using Lagrange interpolation find a rational function $r \in \mathcal{H}^\infty$ such that

$$\frac{r - n_1 x}{m_d} \in \mathcal{H}^\infty .$$

Blaschke Product: Construct a Blaschke product $m_w(z)$ whose zeros are the $1/\overline{p_i}$'s and the $1/\overline{z_j}$'s, where the p_i's are the poles of $w_1 n_2$ and the z_j's are the zeros of g.

Problem Data: Define

$$\begin{aligned} w_{0,\rho} &:= f_\rho^{-1} m_w w_1 w_1^* n_2^* / g^*, \\ \widehat{w}_{0,\rho} &:= f_\rho^{-1} g r, \\ m_1 &:= m_n m_w \ , \\ m_2 &:= m_d \ , \\ m &:= m_1 m_2 \ . \end{aligned}$$

Let us now define three polynomials b_ρ, c_ρ, k_ρ from the problem data:

$$\begin{aligned} w_{0,\rho}(z) &=: b_\rho(z)/k_\rho(z) \ , \\ \widehat{w}_{0,\rho}(z) &=: c_\rho(z)/k_\rho(z). \end{aligned}$$

Suppose that

$$\begin{aligned} b_\rho(z) &= b_0 + b_1 z + \cdots + b_n z^n \ , \\ c_\rho(z) &= c_0 + c_1 z + \cdots + c_n z^n \ , \\ k_\rho(z) &= k_0 + k_1 z + \cdots + k_n z^n \ , \end{aligned}$$

where $k_n \neq 0$ and $k^{-1} \in \mathcal{H}^\infty$. Note that the coefficients b_i's, c_i's and k_i's depend on ρ.

Matricial Formulae:

Below we will show that (5.15) is equivalent to (5.19), and we will give a matricial formulae for $R_\rho'(z)$ which appears in (5.19). For this purpose we compute the left hand side and the right hand side of (5.15) separately, and then we combine these terms in one equation to obtain (5.19).

(i) The left hand side of (5.15): This term can be computed from equations (4.22–4.24) as follows:

$$\Big(k_\rho(\mathbf{T})^* k_\rho(\mathbf{T}) - b_\rho(\mathbf{T})^* b_\rho(\mathbf{T})\Big) y' =$$
$$z^{-n}\left(\chi_\rho(z) y'(z) + R_-(z)\phi_- + R_+(z)\phi_+\right) \tag{5.28}$$

where

$$\chi_\rho(z) = z^n (k_\rho(z^{-1}) k_\rho(z) - b_\rho(z^{-1}) b_\rho(z)),$$

$\phi_- := [\phi_{-n}^* \ \cdots \ \phi_{-1}^*]^*$, $\phi_+ := [\phi_0^* \ \cdots \ \phi_{n-1}^*]^*$, and $R_-(z)$ and $R_+(z)$ are exactly as defined in (4.23) and (4.24), respectively. Note that the ϕ_{-i}'s come from the power series expansion of $(m^* y')(z) = \sum_{k=1}^{\infty} \phi_{-k} z^{-k}$, and the ϕ_i's come from the expansion $y'(z) = \sum_{k=0}^{\infty} \phi_k z^k$.

Let us recall the matricial formulae for $R_-(z)$ and $R_+(z)$. We first defined

$$\begin{aligned} V_+(z) &:= [1 \ \ z \ \ \ldots \ \ z^{n-1}], \\ V_-(z) &:= [z^{-n} \ \ \ldots \ \ z^{-1}], \end{aligned}$$

$$\mathcal{B} := \begin{bmatrix} b_n & \ldots & b_1 \\ 0 & \ddots & \vdots \\ 0 & 0 & b_n \end{bmatrix}, \qquad \mathcal{K} := \begin{bmatrix} k_n & \ldots & k_1 \\ 0 & \ddots & \vdots \\ 0 & 0 & k_n \end{bmatrix},$$

$$\mathcal{M} := \begin{bmatrix} m_0 & 0 & 0 \\ \vdots & \ddots & 0 \\ m_{n-1} & \cdots & m_0 \end{bmatrix}, \quad \mathcal{X} := \begin{bmatrix} \chi_{-n} & 0 & 0 \\ \vdots & \ddots & 0 \\ \chi_{-1} & \cdots & \chi_{-n} \end{bmatrix},$$

where $m(z) =: m_0 + m_1 z + m_2 z^2 + \cdots$. Then, we defined

$$\mathcal{L} := \mathcal{B}^* \mathcal{M} \mathcal{B} - \mathcal{K}^* \mathcal{M} \mathcal{K},$$

$$\tilde{b}(z) := z^n b(z^{-1}), \quad \tilde{k}(z) := z^n k(z^{-1}),$$

and

$$\begin{aligned} R_-(z) &:= V_+(z)\left(m(z)(\tilde{b}(z)\mathcal{B} - \rho^2 \tilde{k}(z)\mathcal{K}) - \mathcal{L}\right), \\ R_+(z) &:= V_+(z)\mathcal{X}. \end{aligned}$$

(ii) The right hand side of (5.15): There are three terms to be computed here:

$$b_\rho(\mathbf{T})^* c_\rho(\mathbf{T}) m_1(\mathbf{T}) y',$$
$$c_\rho(\mathbf{T})^* m_1(\mathbf{T})^* b_\rho(\mathbf{T}) y',$$
$$c_\rho(\mathbf{T})^* m_1(\mathbf{T})^* m_1(\mathbf{T}) c_\rho(\mathbf{T}) y'.$$

Similar to the computations of Chapter 4, we can express the above terms using certain matrices and the entries of Φ. The $n \times n$ upper triangular matrix $\mathcal{C}$ is defined similarly to the definitions of $\mathcal{B}$ and $\mathcal{K}$. Again similar to $\mathcal{M}$ we define the $n \times n$ lower triangular matrices $\mathcal{M}_1$ and $\mathcal{M}_2$ from the power series coefficients of $m_1(z)$ and $m_2(z)$, respectively. Suppose that the zeros $a_1, \ldots, a_\ell$ of m_2 are distinct and none of them is at the origin. Recall that $f_1, \ldots, f_\ell$ form a basis for $\mathcal{H}(m_2)$, where $f_i(z) = (1 - \overline{a_i} z)^{-1}$. Define the $1 \times \ell$ vector

$$F(z) = [f_1(z) \quad \ldots \quad f_\ell(z)],$$

and the $\ell \times \ell$ Pick matrix

$$\Lambda = \begin{bmatrix} F(a_1) \\ \vdots \\ F(a_\ell) \end{bmatrix}.$$

We will also use the $n \times n$ upper triangular matrices

$$\widehat{\mathcal{B}} = \begin{bmatrix} b_0 & \cdots & b_{(n-1)} \\ 0 & \ddots & \vdots \\ 0 & 0 & b_0 \end{bmatrix} \quad \widehat{\mathcal{C}} = \begin{bmatrix} c_0 & \cdots & c_{(n-1)} \\ 0 & \ddots & \vdots \\ 0 & 0 & c_0 \end{bmatrix},$$

the $\ell \times n$ Vandermonde matrix

$$\mathcal{A}_+ = \begin{bmatrix} V_+(a_1) \\ \vdots \\ V_+(a_\ell) \end{bmatrix},$$

and the $\ell \times \ell$ diagonal matrix

$$\mathcal{D}_C = \begin{bmatrix} c_\rho(a_1) & 0 & 0 \\ 0 & \ddots & 0 \\ 0 & 0 & c_\rho(a_\ell) \end{bmatrix}.$$

Recall from Chapter 5 that y' has an orthogonal decomposition of the form $y' = u + m_1 v$, where $u \in \mathcal{H}(m_1)$ and $v \in \mathcal{H}(m_2)$ with

$$v(z) = F(z)\alpha$$

and $\alpha = [\alpha_1^* \ \cdots \ \alpha_\ell^*]^*$. We have also defined $\beta_i := y'(a_i)$, accordingly we set $\beta = [\beta_1^* \ \cdots \ \beta_\ell^*]^*$.
With this notation we can show that

$$(b_\rho(\mathbf{T})^* c_\rho(\mathbf{T}) m_1(\mathbf{T}) y')(z) = \left(b_\rho(z^{-1})F(z)m_1(z) - V_-(z)\mathcal{B}^*\mathcal{M}_1\mathcal{A}_+^*\right)\mathcal{D}_C\Lambda^{-1}\beta. \quad (5.29)$$

Similarly, we have

$$(c_\rho(\mathbf{T})^* m_1(\mathbf{T})^* b_\rho(\mathbf{T}) y')(z) = \left(c_\rho(z^{-1})b_\rho(z)F(z) - V_-(z)\mathcal{C}^*\widehat{\mathcal{B}}^*\mathcal{A}_+^*\right)\alpha + \left(V_+(z)\mathcal{B}\widehat{\mathcal{C}} - c_\rho(z^{-1})m_2(z)V_+(z)\mathcal{B} + V_-(z)\mathcal{C}^*\mathcal{M}_2\mathcal{B}\right)\phi_- \quad (5.30)$$

and

$$(c_\rho(\mathbf{T})^* m_1(\mathbf{T})^* m_1(\mathbf{T}) c_\rho(\mathbf{T}) y')(z) = \left(c_\rho(z^{-1})F(z) - V_-(z)\mathcal{C}^*\mathcal{A}_+^*\right)\mathcal{D}_C\Lambda^{-1}\beta. \quad (5.31)$$

(iii) Equivalent form of (5.15): Note that the terms in the right hand side of (5.15) are in the form

$$R_1(z)\Phi := (b_\rho(\mathbf{T})^* c_\rho(\mathbf{T}) m_1(\mathbf{T}) y')(z)\,, \quad (5.32)$$
$$R_2(z)\Phi := (c_\rho(\mathbf{T})^* m_1(\mathbf{T})^* b_\rho(\mathbf{T}) y')(z)\,, \quad (5.33)$$
$$R_3(z)\Phi := (c_\rho(\mathbf{T})^* m_1(\mathbf{T})^* m_1(\mathbf{T}) c_\rho(\mathbf{T}) y')(z), \quad (5.34)$$

where R_1, R_2 and R_3 are $1 \times 2(n+\ell)$ vector-valued functions defined by rearranging the right hand sides of (5.29–5.31). Also, the left hand side of (5.15) is of the form

$$z^{-n}\chi_\rho(z)y'(z) + R_0(z)\Phi \ , \tag{5.35}$$

where R_0 is a $1 \times 2(n+\ell)$ vector-valued function obtained from $R_+(z)$ and $R_-(z)$ defined above.

In summary, we see that the singular value/singular vector equation (5.15) is equivalent to

$$z^{-n}\chi_\rho(z)y'(z) = R'_\rho(z)\Phi, \tag{5.36}$$

where $R'_\rho(z) := -R_1(z) - R_2(z) + R_3(z) - R_0(z)$.

Let $z_1, \ldots, z_{2n}$ denote the roots of $\chi_\rho(z) = 0$. Then under Assumptions 4.1 and 4.2 of Chapter 4 we have the following necessary conditions for a non-zero $y' \in \mathcal{H}(m)$ to satisfy (5.15):

$$R'_\rho(z_i)\Phi = 0 \qquad \text{for all} \quad i = 1, \ldots, 2n. \tag{5.37}$$

By definition, we have the following ℓ necessary conditions:

$$a_i^{-n}\chi_\rho(a_i)\beta_i = R'_\rho(a_i)\Phi \quad \text{for all} \quad i = 1, \ldots, \ell. \tag{5.38}$$

This last set of ℓ necessary conditions are obtained by taking the orthogonal projection of (5.36) onto $m_1\mathcal{H}^2(\mathbf{D})$. Note that we have

$$\mathbf{P}_+ m_1^* z^{-n}\chi_\rho(z)y' = z^{-n}\chi_\rho(z)F(z)\alpha + V_+(z)\mathcal{X}^*\phi_- - V_-(z)\mathcal{X}\mathcal{A}_+^*\alpha. \tag{5.39}$$

Hence defining $\mathbf{P}_+ m_1^* R'_\rho(z)\Phi =: R_4(z)\Phi$ and combining (5.39) with R_4 we see that

$$z^{-n}\chi_\rho(z)F(z)\alpha = R'_4(z)\Phi, \tag{5.40}$$

for an explicitly computable $1 \times 2(n+\ell)$ function $R'_4(z)$. Evaluating (5.40) at the zeros of m_2 we have

$$a_i^{-n}\chi_\rho(a_i)F(a_i)\alpha = R'_4(a_i)\Phi \quad \text{for all} \quad i = 1, \ldots, \ell. \tag{5.41}$$

To summarize, we have obtained $2(n+\ell)$ equations (5.37), (5.38) and (5.41), in terms of the $2(n+\ell)\times 1$ unknown vector Φ, as necessary conditions for $y' \in \mathcal{H}(m)$ to satisfy (5.15). In [79], it was shown that these equations also constitute a set of sufficient conditions for the existence of a non-zero $y' \in \mathcal{H}(m)$ satisfying (5.15). Thus we conclude that 1 is a singular value of the Sarason operator $w_\rho(\mathbf{T})$ if and only if there exists a non-zero Φ satisfying the $2(n+\ell)$ equations (5.37), (5.38), (5.41). The largest ρ satisfying these conditions gives us the two block $\mathcal{H}^\infty$ optimal performance γ_{opt}. The computation of γ_{opt} from the above equations can be shown to be numerically feasible; see [103] for a detailed discussion on this issue.

The optimal controller can be constructed from γ_{opt} and the above $2(n+\ell)$ equations. Here we briefly outline the procedure, for the details see [80]. First suppose that we have a non-zero Φ_o satisfying the above $2(n+\ell)$ equations for $\rho = \gamma_{opt}$. Then, the maximal singular vector y_o' can be obtained from

$$y_o'(z) = \frac{R'_{\gamma_{opt}}(z)\Phi_o}{z^{-n}\chi_{\gamma_{opt}}(z)}. \tag{5.42}$$

Similar to Theorem 17, it is easy to see from Sarason's theorem that

$$(w_{\gamma_{opt}} - mq_3^o)(z) = \frac{(w_{\gamma_{opt}}(\mathbf{T})y_o')(z)}{y_o'(z)}, \tag{5.43}$$

where $q_3^o = f_{\gamma_{opt}}^{-1} q_3^{opt}$ and q_3^{opt} is the optimal interpolant in the two block $\mathcal{H}^\infty$ optimal control problem definition (5.7). On the other hand, we know that the optimal controller can be computed from the optimal q, which can be computed from the optimal q_3 via transformations q_1, and q_2, as shown in Chapter 5. Therefore, by solving q_3^o from (5.43) we can obtain the optimal controller. Hence, in order to find the optimal controller all we need to do is to compute the right hand side of (5.43). Note that

$$\frac{(w_{\gamma_{opt}}(\mathbf{T})y_o')(z)}{y_o'(z)} = \frac{(b_{\gamma_{opt}}(\mathbf{T}) - c_{\gamma_{opt}}(\mathbf{T})m_1(\mathbf{T}))y_o'}{k_{\gamma_{opt}}(\mathbf{T})y_o'}$$

Using the notation defined above, this expression can be computed in terms of the entries of Φ_o as follows

$$\begin{aligned} b_{\gamma_{opt}}(\mathbf{T})y_o' &= b_{\gamma_{opt}}(z)y_o'(z) - m(z)V_+(z)\mathcal{B}\phi_-^o \\ k_{\gamma_{opt}}(\mathbf{T})y_o' &= k_{\gamma_{opt}}(z)y_o'(z) - m(z)V_+(z)\mathcal{K}\phi_-^o \\ b_{\gamma_{opt}}(\mathbf{T})m_1(\mathbf{T})y_o' &= m_1(z)F(z)\mathcal{D}_C\Lambda^{-1}\beta^o. \end{aligned}$$

These formulae along with $y_o'(z)$ expression given in (5.42) determine q_3^o, which then determines the optimal controller.

Chapter 6

Suboptimal $\mathcal{H}^\infty$ Controller Design

From the results of Chapter 5, it is easy to see that in general the optimal $\mathcal{H}^\infty$ controller for an infinite dimensional plant is itself infinite dimensional. There are several implementation problems with such controllers. For example, in principle one needs infinite memory in the analog or digital computer realizing an infinite dimensional controller. Another (perhaps more serious) problem is that when the dimension of the controller is large, real time computation of the command signal can take a significant amount of time. This computational time delay in the controller is not modeled in the original $\mathcal{H}^\infty$ control problem definition. If this time delay is too large, it will affect the system performance. Therefore, it is important to find reduced order (finite dimensional) controllers whose $\mathcal{H}^\infty$ performance is close to the optimum, γ_{opt}. We will call such controllers *suboptimal.* Note that here the performance of a controller is defined in the $\mathcal{H}^\infty$ sense as in Chapter 3, i.e.,

$$\gamma(C) = \left\| \begin{bmatrix} W_1(1+PC)^{-1} \\ W_2PC(1+PC)^{-1} \end{bmatrix} \right\|_\infty . \tag{6.1}$$

There are several ways to obtain finite dimensional suboptimal $\mathcal{H}^\infty$ controllers for a given infinite dimensional plant, P. These methods can be categorized into two: direct design, and indirect design. The direct design methods can be described as follows:

D1: Given a desired upper bound (say N) for the controller dimension, find the best controller C such that $\gamma(C)$ is minimum among all Nth order controllers stabilizing the closed loop system $[C, P]$.

D2: Given a desired performance bound $\gamma_d > \gamma_{opt}$, find (if it exists) a finite dimensional, and preferably of minimal dimension, controller C such that $\gamma(C) \leq \gamma_d$.

D3: Modify the performance criterion, so that the dimension of the controller enters into the cost function, directly or indirectly.

We claim that any other direct design method can be reformulated as either D1, D2 or D3. To best of our knowledge, the above problems (in the generality posed) are presently open. An important step in D2 is to parametrize all controllers achieving a performance level $\gamma(C) \leq \gamma_d$. Then, among all such controllers we may try to find the finite dimensional, and minimal dimensional ones. In Section 6.2 we will give a parametrization of all suboptimal $\mathcal{H}^\infty$ controllers for possibly unstable infinite dimensional SISO plants. However, it is still very difficult to identify the finite dimensional controllers in this characterization.

There are two basic indirect design methods which we now describe.

I1: Given a desired controller order N, approximate the original plant P so that we have an Nth order plant model P_N. Then, find the optimal controller $\widehat{C}_N$ corresponding to P_N. Check that $[\widehat{C}_N, P]$ is stable and $\gamma(\widehat{C}_N)$ is as desired. If we are not satisfied with the performance of $\widehat{C}_N$, increase N and repeat the procedure until a satisfactory result is obtained.

I2: Find the optimal infinite dimensional controller C_{opt} for the original plant P. Then, approximate C_{opt} by an Nth order controller C_N. Check that $[C_N, P]$ is stable and $\gamma(C_N)$ is as desired. If we are not satisfied with the performance of C_N, increase N and repeat the procedure until a satisfactory result is obtained.

The methods I1 and I2 are well studied in the literature, for example see [12], [71], [84], [87], [112] for I1, and [77] for I2. In Section 6.1

below, we will present some robustness and convergence results from [77] concerning the method I2. Note however that both I1 and I2 rely on approximations of an infinite dimensional system by a finite dimensional one. Therefore, the results obtained in indirect methods depend on the success of the approximation methods used. Although there are some important results and computer programs on approximation of infinite dimensional systems, it is still a current research topic; see e.g., [45], [46], [49], [53], [69], [85], [113].

6.1 Approximation of the optimal controller

As we have seen in Chapter 3, solutions to robust control problems (under additive or multiplicative uncertainty) related to infinite dimensional *unstable* systems, with finitely many unstable modes, require controllers with finitely many closed right half plane poles. However, there is no *a priori* guarantee for this to happen. Also, as briefly discussed above, for practical implementation purposes we may want to obtain a finite dimensional controller to start with. Therefore, one may want to approximate the optimal controller $C_{opt} = (X + DQ_{opt})/(Y - NQ_{opt})$, where $P = N/D$, and $N, X, D, Y \in \mathcal{H}^\infty$ satisfy the corresponding Bezout identity. An obvious way to do this is to approximate each infinite dimensional term in C_{opt} (i.e. Y, N, Q_{opt}) separately, and then combine the approximations. On the other hand, even if such an approximation is possible, we must make sure that it will not destabilize the closed loop system, and will yield a performance γ close to γ_{opt}. This issue will be discussed now. This section is primarily based on the results reported in [77].

6.1.1 Outline of the approximation method

Let us consider the two block $\mathcal{H}^\infty$ control problem

$$\gamma_{opt} = \inf_{[C,P] \text{ stable}} \left\| \begin{bmatrix} W_1(1+PC)^{-1} \\ W_2C(1+PC)^{-1} \end{bmatrix} \right\|_\infty , \tag{6.2}$$

where $W_1 \in \mathcal{H}^\infty$ is rational, $W_2 = W_a$ (the additive plant uncertainty as defined in Chapter 3) satisfies Assumption 3.2, and $P = N/D$, $(N = M_n N_1 N_2,\ D = M_d)$ is the nominal plant, satisfying Assumption 3.1. Recall the structure of the optimal controller

$$C_{opt}(s) = \frac{X(s) + D(s)Q_{opt}(s)}{Y(s) - N(s)Q_{opt}(s)},$$

where $X, Y \in \mathcal{H}^\infty$ satisfy (3.3), and $Q_{opt} \in \mathcal{H}^\infty$ is the optimal solution of the two block $\mathcal{H}^\infty$ problem. Note that when the plant has finitely many unstable modes we can choose $D(s)$ and $X(s)$ as rational functions. Therefore, the infinite dimensional parts of $C_{opt}(s)$ are $N(s)$, $Y(s)$, $Q_{opt}(s)$, and a finite dimensional controller can be obtained by simply replacing these irrational functions by some k-th order approximating rational $\mathcal{H}^\infty$ functions $N_k(s)$, $Y_k(s)$, $Q_k(s)$

$$C_k(s) = \frac{X(s) + D(s)Q_k(s)}{Y_k(s) - N_k(s)Q_k(s)}, \tag{6.3}$$

(here by a slight abuse of notation we use C_k for the finite dimensional controller (6.3) whose order is larger than k). The controller C_k can be expressed as

$$C_k = \frac{(X + DQ_{opt}) + \Delta_k^n}{(Y - NQ_{opt}) + \Delta_k^d}, \tag{6.4}$$

where

$$\Delta_k^n = D(Q_k - Q_{opt}) \tag{6.5}$$

and

$$\Delta_k^d = (Y_k - Y) - N_k(Q_k - Q_{opt}) - Q_{opt}(N_k - N). \tag{6.6}$$

Following [42], we will say that C_k converges to C_{opt} in the *gap metric* if

$$\|[\Delta_k^d \quad \Delta_k^n]\|_\infty \to 0 \text{ as } k \to \infty .$$

We will see that if $\|[\Delta_k^d \;\; \Delta_k^n]\|_\infty$ is "sufficiently small," then C_k stabilizes P, and the performance of C_k

$$\gamma_k := \left\| \begin{bmatrix} W_1(1+PC_k)^{-1} \\ W_2C_k(1+PC_k)^{-1} \end{bmatrix} \right\|_\infty , \tag{6.7}$$

is "close" to the optimal performance γ_{opt}.

6.1.2 Convergence results

Obviously the convergence of $\|\Delta_k^d\|_\infty$ and $\|\Delta_k^n\|_\infty$ to zero depend on whether the infinite dimensional parts Y, N, Q_{opt} are uniformly approximable (in $\mathcal{H}^\infty$) or not. We know that [56] if $N(j\omega)$ is uniformly continuous on the extended imaginary axis $j\mathbb{R}_e$, then N and Y are uniformly approximable by rational functions in $\mathcal{H}^\infty$, i.e. for every $\epsilon > 0$ there exist rational $\mathcal{H}^\infty$ functions N_k and Y_k such that $\|N - N_k\|_\infty < \epsilon$ and $\|Y - Y_k\|_\infty < \epsilon$. On the other hand, Q_{opt} depends on N, D and the weights W_1 and W_2. So, uniform continuity of $Q_{opt}(j\omega)$ depends on the plant and the weights. In [77], certain conditions are given for Q_{opt} to be uniformly approximable by rational functions. The precise statements of these conditions are given below.

Lemma 9 *([77]) Consider the two block $\mathcal{H}^\infty$ mixed sensitivity minimization problem (6.2), with the plant $P = N/D$, satisfying Assumption 3.1, and the weights $W_1 = W_d^{-1}$ and $W_2 = W_a$, satisfying Assumption 3.2. Suppose that $N(j\omega)$ is continuous on $j\mathbb{R}_e$, $N \in \mathcal{H}^2$, and $\frac{d}{ds}N \in \mathcal{H}^1$. Further assume that either (i) N_1 is rational, or (ii) $N_1(j\omega)$ and $M_n(j\omega)N_2(j\omega)$ are continuous on $j\mathbb{R}_e$, and $N_1, (M_nN_2) \in \mathcal{H}^2$, with $\frac{d}{ds}N_1, \frac{d}{ds}(M_nN_2) \in \mathcal{H}^1$. Then, the optimal controller C_{opt} can be uniformly approximated, in the gap metric, by rational functions C_k of the form (6.4):*

$$C_k = \frac{X + DQ_k}{Y_k - N_kQ_k},$$

where Y_k, N_k and Q_k are rational (kth order) approximations of Y, N and Q_{opt} respectively, such that

$$\|Y - Y_k\|_\infty \to 0 \quad \text{as} \quad k \to \infty,$$

$$\|N - N_k\|_\infty \to 0 \quad \text{as} \quad k \to \infty,$$

$$\|Q_{opt} - Q_k\|_\infty \to 0 \quad \text{as} \quad k \to \infty.$$

Proof. For the case (i), where N_1 is rational, the result is given in [77], which uses certain facts from [53] and [86]. If N_1 is not rational, but (ii) is satisfied, then, by arguments similar to the ones used in Lemmas 3.1 and 3.2 of [77], we can still show that Y, N and Q_{opt} are uniformly approximable by rational functions in $\mathcal{H}^\infty$. Uniform approximability of Y and N can be deduced from Theorem 2.12 of [53]. Also by the same theorem we can show that if the optimal solution $Q_{3,opt}$ of (5.7) is uniformly approximable by rational functions in $\mathcal{H}^\infty$ then so is Q_{opt}. In order to show uniform approximability of $Q_{3,opt}$ it is sufficient to check that $m^* w_\rho$ of (5.12) has an absolutely summable power series expansion, see [86] pp. 48–53; and this also is guaranteed by Theorem 2.12 of [53], provided N_1 and $M_n N_2$ satisfy the conditions stated in the lemma. □

Now that we have convergence of C_k to C_{opt}, the closed loop system $[C_k, P]$ stability is established as follows.

Theorem 22 *([77]) Assume that the hypotheses of Lemma 9 hold. Then, there exists a sufficiently large number K such that the closed loop system $[C_k, P]$ is stable for all $k \geq K$.*

Proof. First set $C^n_{opt} := (X + DQ_{opt})$, $C^d_{opt} := (Y - NQ_{opt})$, and define their spectral factor $G_c, G_c^{-1} \in \mathcal{H}^\infty$

$$G_c^* G_c := C^{n*}_{opt} C^n_{opt} + C^{d*}_{opt} C^d_{opt}. \tag{6.8}$$

The fact that $G_c^{-1} \in \mathcal{H}^\infty$ comes from strong coprimeness of C_{opt}^n and C_{opt}^d, which is guaranteed by stability of the closed loop system $[C_{opt}, P]$ (cf. [91]). Now, since P is stabilizable and C_{opt} stabilizes P, there is, (cf. [42]), a positive number $b_{P,C_{opt}} > 0$, such that the closed loop system $[C_k, P]$ is stable if

$$\|G_c^{-1} \begin{bmatrix} \Delta_k^n \\ \Delta_k^d \end{bmatrix} \|_\infty < b_{P,C_{opt}}, \tag{6.9}$$

where G_c, Δ_k^n and Δ_k^d are defined by (6.8), (6.5), (6.6). On the other hand by Lemma 9

$$\varepsilon(k) := \|G_c^{-1} \begin{bmatrix} \Delta_k^n \\ \Delta_k^d \end{bmatrix} \|_\infty \to 0 \quad \text{as} \quad k \to \infty. \tag{6.10}$$

This concludes the proof. □

It is important to note that (see [42]) the quantity $b_{P,C_{opt}}$ can be computed from the formula

$$b_{P,C_{opt}}^{-1} = \| \begin{bmatrix} 1 \\ P \end{bmatrix} (1 + PC_{opt})^{-1} [1 \quad C_{opt}] \|_\infty,$$

which is equal to

$$b_{P,C_{opt}}^{-1} = \left\| G_p G_p^{*-1} \begin{bmatrix} (N^*Y - D^*X) - G_p G_p^* Q_{opt} \\ 1 \end{bmatrix} \right\|_\infty, \tag{6.11}$$

where $G_p, G_p^{-1} \in \mathcal{H}^\infty$, and

$$G_p G_p^* = NN^* + DD^*,$$

see [50]. Therefore, from (6.11) we can determine how small $\varepsilon(k)$ in (6.10) should be to guarantee the stability of the feedback system $[C_k, P]$. This also determines how large K should be in Theorem 6.2.

An important requirement in the approximation scheme C_k is that the performance γ_k of the system $[C_k, P]$, should be close to the optimum γ_{opt}. The next result shows that C_k, obtained from the approximation method proposed above satisfies this requirement.

Theorem 23 *Assume that the hypotheses of Lemma 9 hold. Then,*

$$\gamma_k \longrightarrow \gamma_{opt} \quad \text{as} \quad k \to \infty. \tag{6.12}$$

Proof. First define $C_k^n := X + DQ_k$, $C_k^d := Y_k - N_k Q_k$, and $S_{opt} := (1 + PC_{opt})^{-1}$. Then, one can re-write γ_k as

$$\gamma_k = \left\| \begin{bmatrix} W_1 D C_k^d (DC_k^d + NC_k^n)^{-1} \\ W_2 D C_k^n (DC_k^d + NC_k^n)^{-1} \end{bmatrix} \right\|_\infty .$$

It is easy to see that

$$\gamma_k = \left\| \begin{bmatrix} W_1 S_{opt} + W_1 D \Delta_k^d \\ W_2 C_{opt} S_{opt} + W_2 D \Delta_k^n \end{bmatrix} \left(\frac{1}{1 + (D\Delta_k^d + N\Delta_k^n)} \right) \right\|_\infty .$$

Recall that by Assumption 3.1 (Chapter 3), $D = m_d$ is inner, so we have

$$\gamma_k \le \left(\gamma_{opt} + \left\| \begin{bmatrix} W_1 \Delta_k^d \\ W_2 \Delta_k^n \end{bmatrix} \right\|_\infty \right) \left(\frac{1}{1 - \|(D\Delta_k^d + N\Delta_k^n)\|_\infty} \right). \tag{6.13}$$

On the other hand, by Lemma 9, $\|\Delta_k^n\|_\infty \to 0$, and $\|\Delta_k^d\|_\infty \to 0$, as $k \to \infty$. Since W_1, W_2, N and D are in $\mathcal{H}^\infty$, we conclude that $\gamma_k \to \gamma_{opt}$ as $k \to \infty$. □

This theorem gives an explicit bound (6.13) for the performance degradation

$$\gamma_k - \gamma_{opt} \le \|(D\Delta_k^d + N\Delta_k^n)\|_\infty \gamma_k + \left\| \begin{bmatrix} W_1 \Delta_k^d \\ W_2 \Delta_k^n \end{bmatrix} \right\|_\infty . \tag{6.14}$$

This bound is expressed in terms of the approximation errors Δ_k^n and Δ_k^d, which depend on the specific methods used in the approximation of infinite dimensional terms Q_{opt}, Y, N.

6.2 Suboptimal controllers via AAK theory

In this section, we present an explicit formula for suboptimal $\mathcal{H}^\infty$ controllers for infinite dimensional plants with a finite number of unstable poles. This section of the book is based on [52, 98, 99, 101]. The $\mathcal{H}^\infty$ suboptimal control problem for *stable* distributed plants was first solved in [35]. Here we will use the Adamjan-Arov-Krein (AAK) theory to characterize all suboptimal $\mathcal{H}^\infty$ controllers. The key step in the solution of this problem is the computation of the action of a certain Hankel operator on the function $e(z) = 1$. The Hankel operator has a factorizable symbol, and it is similar to the Sarason operator defined in Chapter 5. Therefore, in the solution, the approach of [79] (summarized in Chapter 5) plays an important role.

6.2.1 Problem definition

We consider the class of plants defined in Chapter 3:

$$P = \frac{N}{D}$$

where $N \in \mathcal{H}^\infty$ and $D = m_d \in \mathcal{H}^\infty$, satisfy Assumption 3.1.

The following mixed sensitivity reduction problem will be studied: find all controllers in the set

$$\mathcal{C}_\rho = \left\{ C \;:\; [C, P] \text{ stable, and } \quad \left\| \begin{bmatrix} W_1 S \\ W_2 T \end{bmatrix} \right\|_\infty \le \rho \right\} \tag{6.15}$$

where W_1, W_2 are the weights satisfying Assumption 3.2, and $S = (1 + PC)^{-1}$ and $T = 1 - S$.

As shown in Chapter 5, the problem can be reduced to the parametrization of

$$\mathcal{S}_\rho = \left\{ \hat{q} \in \mathcal{H}^\infty \;:\; \left\| \begin{bmatrix} w_0 - \hat{w}_0 m_1 - m\hat{q} \\ g_0 \end{bmatrix} \right\|_\infty \le \rho \right\}, \tag{6.16}$$

where w_0, $\widehat{w}_0$, g_0, m_1, $m = m_1 m_2$ are defined in Chapter 5. Note that

$$\gamma_{opt} := \inf_{\widehat{q}\in\mathcal{H}^\infty} \left\| \begin{bmatrix} w_0 - \widehat{w}_0 m_1 - m\widehat{q} \\ g_0 \end{bmatrix} \right\|_\infty \tag{6.17}$$

which can be computed by the formulae given in Section 5.6.

We assume that $\rho > \gamma_{opt}$ is given, then $\|g_0\|_\infty \leq \gamma_{opt} < \rho$; so there exists an $f_\rho \in \mathcal{H}^\infty$, rational, with $f_\rho^{-1} \in \mathcal{H}^\infty$ such that $f_\rho^* f_\rho = \rho^2 - g_0^* g_0$. Hence (6.16) reduces to

$$\mathcal{S}_\rho = \{f_\rho q_1 \ : \ \|w - mq_1\|_\infty \leq 1 \text{ and } q_1 \in \mathcal{H}^\infty\}, \tag{6.18}$$

where $q_1 := \widehat{q}/f_\rho$ and $w := (w_0 - m_1\widehat{w}_0)f_\rho^{-1}$.

6.2.2 Problem solution via AAK approach

We now present a solution to (6.18) via the AAK theory. We will use the following observation.

Lemma 10 *For w and m defined as before, we have*

$$\inf_{q_1\in\mathcal{H}^\infty} \|w - mq_1\|_\infty < 1 \ .$$

Proof. By Theorem 21 of Chapter 5 there exists a $\widehat{q}_o \in \mathcal{H}^\infty$ such that

$$|w_0 - \widehat{w}_0 m_1 - m\widehat{q}_o|^2 + |g_0|^2 \ = \ \gamma_{opt}^2 \ .$$

Obviously

$$|w_0 - \widehat{w}_0 m_1 - m\widehat{q}_o|^2 \ = \ \gamma_{opt}^2 - \rho^2 + |f_\rho|^2 \ ,$$

$$|w - mq_1^o| \ = \ 1 - \frac{\rho^2 - \gamma_{opt}^2}{|f_\rho|^2},$$

where $q_1^o := f_\rho^{-1}\hat{q}_o$. Since $f_\rho^{-1} \in \mathcal{H}^\infty$ and $\rho > \gamma_{opt}$,

$$\|w - mq_1^o\|_\infty < 1 .$$

Hence

$$\inf_{q_1 \in \mathcal{H}^\infty} \|w - mq_1\|_\infty < 1. \quad \square$$

Before we state the AAK theorem we would like to recall some of the notation from Chapter 2. The space $\mathcal{H}(m)$ is the orthogonal complement of $m\mathcal{H}^2$ in $\mathcal{H}^2$. The Hankel operator with symbol m^*w is denoted by $\mathbf{\Gamma}_{m^*w}$. Let $\mathcal{R}$ be the reflection operator from $\mathcal{L}^2 \ominus \mathcal{H}^2$ to $\mathcal{H}^2$, defined by $(\mathcal{R}f)(z) = z^{-1}f(z^{-1})$. Then, $\mathcal{R}^*$ is from $\mathcal{H}^2$ to $\mathcal{L}^2 \ominus \mathcal{H}^2$ and $(\mathcal{R}^*f)(z) = f(z^{-1})z^{-1}$. Finally, recall that $\mathbf{S}$ denotes the shift operator and $\mathbf{T}$ denotes the compressed shift operator on $\mathcal{H}(m)$. We will use $\mathbf{\Gamma}$ for the matrix representation of $\mathcal{R}\mathbf{\Gamma}_{m^*w}$.

Following [1], we now define $\mathbf{R}_\rho := (\mathbf{I} - \overline{\mathbf{\Gamma}}\mathbf{\Gamma})^{-1}$, $e(z) := 1$, $p := \mathbf{R}_\rho e$, and $q := \mathbf{S}\ \overline{\mathbf{\Gamma}}\ \overline{\mathbf{R}}_\rho e$. Here for an operator $\mathbf{X}$ on $\mathcal{H}^2$ we denote $\overline{\mathbf{X}}$ the operator defined by $\overline{\mathbf{X}}f = \overline{(\mathbf{X}\overline{f})}$, where for a function $f \in \mathcal{H}^2$, $\overline{f}(z) = \overline{f(\overline{z})}$ for $z \in \mathsf{D}$.

Theorem 24 (AAK) *Suppose that* $\|\mathbf{\Gamma}_{m^*w}\| < 1$. *Then, the set of all* $q_1 \in \mathcal{H}^\infty$ *satisfying*

$$\|w - mq_1\|_\infty \leq 1$$

can be obtained from

$$m^*w - q_1 = \frac{u\, p^* + q^*}{p + u\, q} \tag{6.19}$$

where $u \in \mathcal{B}^\infty := \{u \in \mathcal{H}^\infty \ :\ \|u\|_\infty \leq 1\}$, *is the free parameter.*

Note that by Lemma 10 and Nehari's theorem we have $\|\mathbf{\Gamma}_{m^*w}\| < 1$. Therefore, Theorem 24 is applicable to our problem. Hence, the solution of (6.18) amounts to finding p and q in (6.19). We now briefly outline the computation of p and q.

Computation of p

We begin with some observations. As shown in the proof of Theorem 16, $\mathbf{\Gamma}_{m^*w}|_{\mathcal{H}(m)} = m^*w(\mathbf{T})$, and hence

$$w(\mathbf{T})^*w(\mathbf{T}) = \mathbf{\Gamma}^*_{m^*w}\mathbf{\Gamma}_{m^*w}|_{\mathcal{H}(m)} = \mathbf{\Gamma}^*\mathbf{\Gamma}|_{\mathcal{H}(m)}, \tag{6.20}$$

because $\mathbf{\Gamma} = \mathcal{R}\mathbf{\Gamma}_{m^*u}$, and $\mathcal{R}^*\mathcal{R} = \mathbf{I}$. Furthermore, $\mathbf{\Gamma}^* = \overline{\mathbf{\Gamma}}$ because $\mathbf{\Gamma}^T = \mathbf{\Gamma}$.

Recall that by the AAK formulae, we have

$$(\mathbf{I} - \mathbf{\Gamma}^*\mathbf{\Gamma})p = e.$$

Now we can decompose the above equation into two orthogonal parts. We have

$$(\mathbf{I} - \mathbf{\Gamma}^*\mathbf{\Gamma})\mathbf{P}_{m\mathcal{H}^2}p + (\mathbf{I} - \mathbf{\Gamma}^*\mathbf{\Gamma})\mathbf{P}_{\mathcal{H}(m)}p = 1.$$

Since $\mathbf{\Gamma}|_{m\mathcal{H}^2} = 0$, we obtain

$$(\mathbf{I} - \mathbf{\Gamma}^*\mathbf{\Gamma})\mathbf{P}_{m\mathcal{H}^2}p = \mathbf{P}_{m\mathcal{H}^2}p \in m\mathcal{H}^2.$$

Also, by (6.20)

$$(\mathbf{I} - \mathbf{\Gamma}^*\mathbf{\Gamma})\mathbf{P}_{\mathcal{H}(m)}p = (\mathbf{I} - w(\mathbf{T})^*w(\mathbf{T}))\mathbf{P}_{\mathcal{H}(m)}p \in \mathcal{H}(m).$$

Therefore, the $m\mathcal{H}^2$ part of p is

$$\mathbf{P}_{m\mathcal{H}^2}p = \mathbf{P}_{m\mathcal{H}^2}1 = \overline{m(0)}m(z). \tag{6.21}$$

Defining

$$y := \mathbf{P}_{\mathcal{H}(m)}p \quad \text{and} \quad \mu(z) := -\mathbf{P}_{\mathcal{H}(m)}1 = -1 + \overline{m(0)}m(z), \tag{6.22}$$

we obtain

$$(\mathbf{I} - w(\mathbf{T})^*w(\mathbf{T}))y = \mathbf{P}_{H(m)}1 = -\mu(z). \tag{6.23}$$

Note that the equation (6.23) is similar to (5.15), the only difference is the extra term $-\mu(z)$. All the other terms can be computed explicitly, as shown in Chapter 5, and hence it is possible to construct $y(z)$ from the solution of a set of finitely many linear equations. Thus, $p(z)$ can be computed explicitly, and the final answer can be expressed in the form

$$p(z) = R_1(z) + m(z)R_2(z), \tag{6.24}$$

where $R_1(z)$ and $R_2(z)$ are rational functions in $\mathcal{L}^2$.

Computation of q

Recall that $q = \mathbf{S}\,\overline{\mathbf{\Gamma}}\,\overline{\mathbf{R}}_\rho e$, where $e(z) = 1$, and other symbols are as defined before. Note that $\overline{\mathbf{R}}_\rho e = \overline{p}(z)$, because $\mathbf{R}_\rho e = p(z)$. Since $\overline{\mathbf{\Gamma}} = \mathbf{\Gamma}^*_{m^*w}\mathcal{R}^*$, we have $q(z) = z\mathbf{\Gamma}^*_{m^*w}\mathcal{R}^*\overline{p}(z)$, so

$$q(z) = z\mathbf{\Gamma}^*_{m^*w}z^{-1}p^*(z), \quad \text{because} \quad \mathcal{R}^*\overline{p}(z) = z^{-1}\overline{p}(z^{-1}) = z^{-1}p^*(z),$$

where $p^*(z) := \overline{p}(z^{-1})$. Hence

$$q(z) = z\mathbf{P}_+ w(z^{-1})z^{-1}(m(z)R_1^*(z) + R_2^*(z)),$$

where $w = f_\rho^{-1}(w_0 - m_1\widehat{w}_0)$, $m = m_1m_2$, and R_1 and R_2 come from (6.24). The right hand side can be computed explicitly in a way similar to the computations of the projections appearing in Section 5.6. The final result is in the form

$$q(z) = R_3(z) + m(z)R_4(z)\ ,$$

where $R_3, R_4 \in \mathcal{L}^2$ are rational functions. In particular it can be shown that $R_4 = f_\rho^{-1}w_0^*R_1^*$ and $R_2 = f_\rho^{-1}w_0R_3^*$; see [98] for the details. Therefore, the only rational functions we need to compute are R_1 and R_3.

6.2.3 Structure of all $\mathcal{H}^\infty$ controllers

Note that all suboptimal $\mathcal{H}^\infty$ controllers can be computed from q_1 which is of the form

$$q_1 = m^* w - \frac{p^* u + q^*}{p + uq}$$

while $u \in \mathcal{B}^\infty$ is arbitrary, p and q can be computed from R_1 and R_3 as shown above, and m and w come from the problem data, see Chapter 5. After substitution, and some algebra we can obtain all suboptimal $\mathcal{H}^\infty$ controllers from this result. In [98], [101] these steps are performed and a simplified formula is derived for the optimal controller as well as the suboptimal ones. The final result in [101] is given in the s domain, i.e. the controller is expressed in terms of the original s domain problem data. We now summarize the main results of [101] without proofs.

Recall that the problem data consists of the weights $W_1(s)$, $W_2(s)$, and the plant $P(s)$, which admits a coprime-inner/outer factorization of the form $P(s) = M_n(s)N_o(s)/M_d(s)$, where $M_d \in R\mathcal{H}^\infty(\mathbb{C}_+)$, $M_n \in \mathcal{H}^\infty(\mathbb{C}_+)$ are inner and $N_o \in \mathcal{H}^\infty(\mathbb{C}_+)$ is outer.

Let $\alpha_1, \ldots \alpha_\ell \in \mathbb{C}_+$ be the zeros of $M_d(s)$, i.e. the unstable poles of $P(s)$, and $\eta_1, \ldots, \eta_{n_1} \in \overline{\mathbb{C}}_+$, be the poles of $W_1(-s)$ (if η_i has multiplicity k_i then it is assumed to be repeated k_i times in this list); and set

$$E_\rho(s) := \left(\frac{W_1(-s)W_1(s)}{\rho^2} - 1 \right). \tag{6.25}$$

The zeros of $E_\rho(s)$ are denoted by $\beta_1, \ldots, \beta_{2n_1}$, and they are assumed to be distinct. Then, β_i's can be enumerated in such a way that $\beta_1, \ldots, \beta_{n_1}$ are in $\overline{\mathbb{C}}_+$, and $\beta_{n_1+i} = -\beta_i$. Now define

$$F_\rho(s) \ := \ G_\rho(s) \prod_{k=1}^{n_1} \frac{s - \eta_k}{s + \eta_k} \tag{6.26}$$

where $G_\rho \in \mathcal{H}^\infty(\mathbb{C}_+)$ is minimum phase and determined from the spectral factorization

$$G_\rho(s)G_\rho(-s) := \left(1 - \left(\frac{W_2(-s)W_2(s)}{\rho^2} - 1\right) E_\rho(s) \right)^{-1}. \tag{6.27}$$

Then (under standard genericity assumptions, similar to the ones in made in Chapter 4, see [101] for full details) the optimal $\mathcal{H}^\infty$ controller is given by

$$C_{opt}(s) = E_{\gamma_o}(s)m_d(s)\,\frac{N_o(s)^{-1}F_{\gamma_o}(s)L(s)}{1+m_n(s)F_{\gamma_o}(s)L(s)} \tag{6.28}$$

where $L(s) = L_2(s)/L_1(s)$, and $L_1(s), L_2(s)$ are polynomials of degrees less than or equal to $(n_1 + l - 1)$ which satisfy

$$\begin{aligned}
0 &= L_1(\beta_k) + m_n(\beta_k)F_{\gamma_o}(\beta_k)L_2(\beta_k) & k &= 1,\ldots,n_1 & (6.29)\\
0 &= L_1(\alpha_k) + m_n(\alpha_k)F_{\gamma_o}(\alpha_k)L_2(\alpha_k) & k &= 1,\ldots,l & (6.30)\\
0 &= L_2(-\beta_k) + m_n(\beta_k)F_{\gamma_o}(\beta_k)L_1(-\beta_k) & k &= 1,\ldots,n_1 & (6.31)\\
0 &= L_2(-\alpha_k) + m_n(\alpha_k)F_{\gamma_o}(\alpha_k)L_1(-\alpha_k) & k &= 1,\ldots,l. & (6.32)
\end{aligned}$$

Note that (6.29–6.30) correspond to interpolation conditions that the denominator term $(1 + m_n(s)F_{\gamma_o}(s)L(s))$ must cancel the closed right half plane zeros of $E_{\gamma_o}(s)m_d(s)$. This means, in particular, that $m_d(s)$ term in the numerator of $C_{opt}(s)$ does not cancel the unstable poles of the plant. Moreover, since $W_1, (W_2N_o)^{-1} \in \mathcal{H}^\infty(\mathbb{C}_+)$, the term $F_{\gamma_o}N_o^{-1}$ is proper. Also note that (6.29–6.32) constitute $2(n_1 + l)$ linear homogeneous equations in the $2(n_1 + l)$ unknown coefficients of $L_1(s)$ and $L_2(s)$. If γ_o is replaced by a variable, say γ, in equations (6.29–6.32), then a new set of linear homogenous equations is obtained, in terms of $2(n_1 + l)$ unknown coefficients, for each fixed γ. The optimal $\mathcal{H}^\infty$ performance γ_o is the largest value of γ for which there is a non-trivial solution to these $2(n_1+l)$ linear homogenous equations. That is, γ_o can be found by plotting smallest singular values of the matrix representation of these equations, as γ varies in an interval. The largest value of γ for which the plot shows a zero is γ_o; see the example of the next section.

All suboptimal $\mathcal{H}^\infty$ controllers are in the form

$$C_{subopt}(s) = E_\rho(s)m_d(s)\,\frac{N_o(s)^{-1}F_\rho(s)L_U(s)}{1+m_n(s)F_\rho(s)L_U(s)} \tag{6.33}$$

where

$$L_U(s) = \frac{L_2(s) + L_1(-s)U(s)}{L_1(s) + L_2(-s)U(s)} ,$$

with $U \in \mathcal{H}^\infty(\mathbb{C}_+)$ with $\|U\|_\infty \leq 1$, and $L_1(s), L_2(s)$ are polynomials of degree $\leq n_1 + l$ satisfying (6.29-6.32) with γ_o replaced by ρ, and the following two conditions:

$$0 = L_2(-a) + (E_\rho(a) + 1)F_\rho(a)m_n(a)L_1(-a) \qquad (6.34)$$

$$1 = L_1(-a), \qquad (6.35)$$

for some arbitrary $a \in \mathrm{R}$, and $a > 0$. In (6.34,6.35) a is a free parameter. It represents the conformal map parameter as in Chapter 2.3. We assume that a is distinct form β_i's and α_j's. For different values of a one can obtain different parametrizations of the suboptimal controllers; see below for a detailed discussion and an example. Also, (6.35) can be replaced by $L_1(-a) \neq 0$, because L_U depends only on the ratios of these polynomials.

6.2.4 Example

In order to find C_{opt}, one needs to compute γ_o and corresponding $L(s)$. Similarly, all suboptimal controllers can be found by obtaining L_1, L_2 which satisfy (6.29–6.32,6.34,6.35). We now present a delay system example to illustrate the computation procedure.

Let $P(s) = e^{-hs}/(s-1)$, and choose $W_1(s) = 2(s+1)/(10s+1)$, $W_2(s) = 0.2(s+1.1)$. The same $\mathcal{H}^\infty$ optimal control problem has been studied in [20] and [21], with slightly different weights. In this example $M_n(s) = e^{-hs}$, $M_d(s) = (s-1)/(s+1)$ and $N_o(s) = 1/(s+1)$. Set $n_1 = 1$, $\ell = 1$, and $\alpha_1 = 1$. The polynomials $L_1(s)$ and $L_2(s)$ are of the form $L_1(s) =: L_{11}s + L_{10}$, and $L_2(s) =: L_{21}s + L_{20}$. If $h = 0.2$, then from [20] we have $0.2 < \gamma_o < 1.5$. In this range of γ,

$$E_\gamma(s) = \frac{(4-\gamma^2) + (100\gamma^2 - 4)s^2}{\gamma^2(1-100s^2)}$$

has two zeros on the imaginary axis. Only one of them, say $\beta_1 = j\sqrt{\frac{4-\gamma^2}{100\gamma^2-4}}$, is used in (6.29–6.32). Also note that

$$F_\gamma(s) = \frac{\gamma^2(1-10s)}{(s+p_1)(s+p_2)\sqrt{4\gamma^2-0.16}},$$

where $p_1 = \sqrt{b+\sqrt{b^2-c}}$ and $p_2 = \sqrt{b-\sqrt{b^2-c}}$, with

$$b = \frac{8.88\gamma^2 - 0.3536}{8\gamma^2 - 0.32} \quad \text{and} \quad c = \frac{4.0484\gamma^2 - 0.1936}{4\gamma^2 - 0.16}.$$

The set of equations (6.29–6.32) can be written as $\mathcal{M}_\gamma \Psi = 0$, where

$$\Psi := [L_{10},\ L_{11},\ L_{20},\ L_{21}]^T, \qquad \mathcal{M}_\gamma := \begin{bmatrix} \mathcal{M}_{\gamma,1} & \mathcal{M}_{\gamma,2} \end{bmatrix}, \quad \text{and}$$

$$\mathcal{M}_{\gamma,1} := \begin{bmatrix} 1 & \beta_1 \\ 1 & \alpha_1 \\ M_n(\beta_1)F_\gamma(\beta_1) & -\beta_1 M_n(\beta)F_\gamma(\beta_1) \\ M_n(\alpha_1)F_\gamma(\alpha_1) & -\alpha_1 M_n(\alpha_1)F_\gamma(\alpha_1) \end{bmatrix},$$

$$\mathcal{M}_{\gamma,2} := \begin{bmatrix} M_n(\beta_1)F_\gamma(\beta_1) & \beta_1 M_n(\beta_1)F_\gamma(\beta_1) \\ M_n(\alpha_1)F_\gamma(\alpha_1) & \alpha_1 M_n(\alpha_1)F_\gamma(\alpha_1) \\ 1 & -\beta_1 \\ 1 & -\alpha_1 \end{bmatrix}.$$

Figure 6.1 shows that the largest value of γ which makes $\mathcal{M}_\gamma$ singular is $\gamma_o = 0.6819$. Now, a non-zero Ψ_o, which satisfies $\mathcal{M}_{\gamma_o}\Psi_o = 0$, can be easily obtained. The entries of Ψ_o give $L(s) = L_2(s)/L_1(s)$. For the above example we found that $L(s) = (s+0.2129)/(s-0.2129)$.

Let us now consider the suboptimal control problem for the same plant and weights. First, define

$$L_0(s) := L_2(s)/L_1(s), \text{ and } D(s) = L_2(-s)/L_1(s).$$

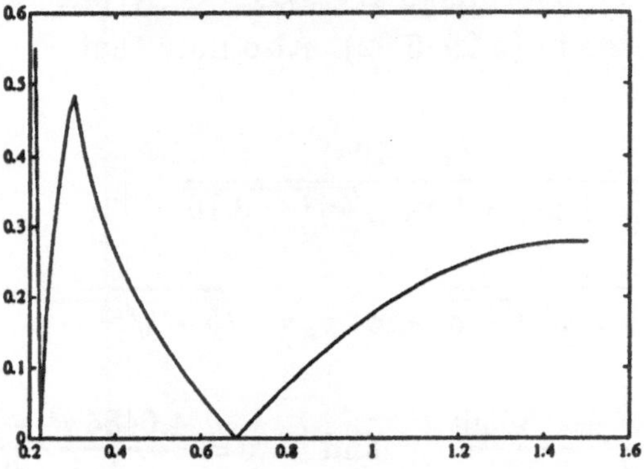

Figure 6.1: $\sigma_{min}(\mathcal{M}_\gamma)$ versus γ

Then L_U, which determines all suboptimal controllers, is given by

$$L_U(s) = L_O(s)\frac{1 + U(s)/D(-s)}{1 + D(s)U(s)}, \quad U \in \mathcal{B}(\mathbb{C}_+).$$

Case 1. If $\rho = 0.7$ and $a = 2$, then we get

$$L_O(s) = \frac{0.96(s+2)(s+0.215)}{(s+1.94)(s-0.21)}, \quad D(s) = \frac{0.96(s-2)(s-0.215)}{(s+1.94)(s-0.21)}.$$

Case 2. If $\rho = 0.7$ and $a = 3$, then we get

$$L_O(s) = \frac{0.94(s+3)(s+0.213)}{(s+2.84)(s-0.21)}, \quad D(s) = \frac{0.94(s-3)(s-0.213)}{(s+2.84)(s-0.21)}.$$

Case 3. If $\rho = 1.0$ and $a = 0.5$, then we get

$$L_O(s) = \frac{0.8(s^2+0.82s+0.18)}{(s+0.73)(s-0.14)}, \quad D(s) = \frac{0.8(s^2-0.82s+0.18)}{(s+0.73)(s-0.14)}.$$

Remarks: Numerical examples given above indicate that when ρ is close to γ_o, $L_O(s)$ is close to $L(s)$ of the optimal case. Moreover, the free part of the controller, i.e. the ratio $\frac{1+U(s)/D(-s)}{1+D(s)U(s)}$, is close to 1 for all $U \in \mathcal{H}^\infty(\mathbb{C}_+)$ with $\|U\|_\infty \leq 1$. This means that, when ρ is close to γ_o, all suboptimal controllers are "close" to the optimal controller. But when ρ is considerably larger than γ_o, the central controller ($U = 0$ case) is significantly different than the optimal controller. Furthermore, in this case the central controller depends on the choice of a; and the range of $\frac{1+U(s)/D(-s)}{1+U(s)D(s)}$ is considerably larger.

Chapter 7

Benchmark Examples

In this chapter we present $\mathcal{H}^\infty$ optimal controller design examples for two different infinite dimensional plants. The first one is a spatially distributed system: a flexible beam model. The second one is a first order unstable system with a time delay. We will use the theory described in Chapters 4, 5 and 6 to solve $\mathcal{H}^\infty$ optimal control problems associated with these plants. The flexible beam example is based on [66], [67], [68], while the second example employs formulae derived from [20, 101]. For more complicated delay system examples see [100] and [104].

7.1 $\mathcal{H}^\infty$ control of a flexible beam

Here we consider the Euler-Bernoulli model of a free-free beam with Kelvin-Voigt damping. We will first derive the transfer functions of two different actuator/sensor configurations. In the first configuration the displacement of one end of the beam is to be controlled by a force applied to the same end of the beam. In the second configuration the displacement of one end of the beam is to be controlled by a force applied to the opposite end of the beam. Then, we solve the weighted mixed sensitivity minimization problem by computing the optimal performance and the corresponding optimal controller. It is possible to obtain transfer functions for other types of beam configurations, (e.g., a beam with hinged ends, cantilevered beam, free-free beam with point force and moment applied to the middle of the beam). In [66] and [68] it was shown that the configurations we are going to study here have the essential features (as far as $\mathcal{H}^\infty$ control is concerned) of these other beam configurations.

We represent the transcendental beam transfer functions in the frequency domain as infinite products of second order terms. The infinite product representations facilitate coprime factorizations and inner/outer factorizations of the plants. As we have seen in the previous chapters, these factorizations are crucial in solving $\mathcal{H}^\infty$ problems.

7.1.1 Beam transfer functions

The basic Euler-Bernoulli model of a flexible beam is given by

$$\rho \frac{\partial^2 w}{\partial t^2} + \frac{\partial^2}{\partial x^2}\left(EI \frac{\partial^2 w}{\partial x^2}\right) = 0.$$

Here $w(x,t)$ denotes the deflection of the beam at time $t \geq 0$ and location x along the beam, $\rho(x)$ denotes the mass density per unit length of the beam, and $EI(x)$ denotes the second moment of the modulus of elasticity about the elastic axis. In this model, it is assumed that $w(x,t)$ is smooth and that no energy is lost or gained internally within the beam.

We will include a damping term to account for beam vibration attenuation due to dissipative mechanisms internal to the beam. There are several damping models for flexible beams; see e.g. [88]. Here we will consider the Kelvin-Voigt damping mechanism, where one assumes that the dissipative forces are obtained from the velocity in the same way that the restoring forces are obtained from the displacement. The dynamics for an Euler-Bernoulli beam with Kelvin-Voigt damping are described by

$$\rho \frac{\partial^2 w}{\partial t^2} + 2\alpha \frac{\partial^2}{\partial x^2}\left(EI \frac{\partial^3 w}{\partial x^2 \partial t}\right) + \frac{\partial^2}{\partial x^2}\left(EI \frac{\partial^2 w}{\partial x^2}\right) = 0 \text{ for } \alpha > 0.$$

We will assume that the damping coefficient is $2\alpha = \epsilon > 0$, the length of the beam is 1, and the other constants are $\rho = 1$, and $EI = 1$. Suppose that a transverse force, $-u(t)$, is applied at one end of the beam, say at $x = 1$. The dynamics of the beam ([67], [88]) are described by the linear partial differential equation

$$\frac{\partial^2 w}{\partial t^2} + \epsilon \frac{\partial^5 w}{\partial x^4 \partial t} + \frac{\partial^4 w}{\partial x^4} = 0, \tag{7.1}$$

with boundary conditions

$$\frac{\partial^2 w}{\partial x^2}(0,t) + \epsilon \frac{\partial^3 w}{\partial x^2 \partial t}(0,t) = 0, \quad \frac{\partial^2 w}{\partial x^2}(1,t) + \epsilon \frac{\partial^3 w}{\partial x^2 \partial t}(1,t) = 0, \tag{7.2}$$

$$\frac{\partial^3 w}{\partial x^3}(0,t) + \epsilon \frac{\partial^4 w}{\partial x^3 \partial t}(0,t) = 0, \quad \frac{\partial^3 w}{\partial x^3}(1,t) + \epsilon \frac{\partial^4 w}{\partial x^3 \partial t}(1,t) = u(t).$$

Now consider the following two cases for the measurement of the beam displacement: $y_1(t) = w(1,t)$, where $w(1,t)$ is the deflection at the end of the beam where the force $-u(t)$ is applied, and $y_2(t) = w(0,t)$, where $w(0,t)$ is the deflection at the other end of the beam. In other words, in the first case the actuator and the sensor are collocated, and in the second case they are non-collocated. The transfer functions $P_1(s) = y_1(s)/u(s)$, and $P_2(s) = y_2(s)/u(s)$, where s denotes the Laplace transform variable, are computed in [67] as given below. First define $\beta^4 = \frac{-s^2}{(1+\epsilon s)}$, then

$$P_1 = \frac{w(1,s)}{u(s)} = \frac{1}{(1+\epsilon s)\beta^3}\left(\frac{\sinh\beta\cos\beta - \cosh\beta\sin\beta}{\cos\beta\cosh\beta - 1}\right), \tag{7.3}$$

$$P_2 = \frac{w(0,s)}{u(s)} = \frac{1}{(1+\epsilon s)\beta^3}\left(\frac{\sinh\beta - \sin\beta}{\cos\beta\cosh\beta - 1}\right). \tag{7.4}$$

One can show that both $P_1(s)$ and $P_2(s)$ are meromorphic in $\mathbb{C}\setminus\left\{\frac{-1}{\epsilon}\right\}$ with no poles in the open right half plane. Also, P_1 and P_2 are strictly proper in the sense that $|P_1(s)| \to 0$ and $|P_2(s)| \to 0$ as $|s| \to \infty$ for $Re(s) \geq 0$, see [67] and [68] for the details. Both P_1 and P_2 can be expressed as infinite products of second order terms. These product representations display the poles and zeros of P_1 and P_2. They also facilitate inner/outer factorizations.

Let us define two sequences α_n and ϕ_n from the roots of the following equations:

$$\begin{aligned}\cos\alpha_n \sinh\alpha_n &= \sin\alpha_n \cosh\alpha_n,\\ \cos\phi_n \cosh\phi_n &= 1,\end{aligned}$$

for $\alpha_n, \phi_n > 0$. Assuming the ordering $\phi_j < \phi_k$ and $\alpha_j < \alpha_k$ for $j < k$, one can establish that the locations of the ϕ_n's alternate with the locations of the α_n's. In fact, the ϕ_n's tend to $\pi/2 + n\pi$ and the α_n's tend to $\pi/4 + n\pi$ as $n \to \infty$.

Lemma 11 *([67]) With the above notation we have*

$$P_1(s) = \frac{-4}{s^2}\prod_{n=1}^{\infty}\left(\frac{1+\epsilon s+\frac{s^2}{\alpha_n^4}}{1+\epsilon s+\frac{s^2}{\phi_n^4}}\right), \tag{7.5}$$

$$P_2(s) = \frac{2}{s^2}\prod_{n=1}^{\infty}\left(\frac{1+\epsilon s-\frac{s^2}{4\alpha_n^4}}{1+\epsilon s+\frac{s^2}{\phi_n^4}}\right), \tag{7.6}$$

where the infinite products (7.5) and (7.6) converge everywhere in the closed right half plane and can be written as quotients of $\mathcal{H}^\infty$ functions. A coprime factorization of P_1, over $\mathcal{H}^\infty$, is $P_1 = P_{1o}D^{-1}$ where $D(s) = (\frac{s}{s+1})^2$, and

$$P_{1o}(s) = \frac{-4}{(s+1)^2}\prod_{n=1}^{\infty}\frac{\left(1+\epsilon s+\frac{s^2}{\alpha_n^4}\right)}{\left(1+\epsilon s+\frac{s^2}{\phi_n^4}\right)}.$$

The plant P_2 can be factored as $P_2 = N_2D^{-1}$, where N_2 has an inner/outer factorization $N_2 = BP_{2o}$, where

$$P_{2o}(s) = \frac{2}{(s+1)^2}\prod_{n=1}^{\infty}\frac{\left(1+s\sqrt{\epsilon^2+\frac{1}{\alpha_n^4}}+\frac{s^2}{4\alpha_n^4}\right)}{\left(1+\epsilon s+\frac{s^2}{\phi_n^4}\right)}, \tag{7.7}$$

$$B(s) = \prod_{n=1}^{\infty} \left(\frac{2\alpha_n^4 \left(\epsilon + \sqrt{\epsilon^2 + \frac{1}{\alpha_n^4}}\right) - s}{2\alpha_n^4 \left(\epsilon + \sqrt{\epsilon^2 + \frac{1}{\alpha_n^4}}\right) + s} \right). \tag{7.8}$$

In [67], it was shown that $P_{1o}, P_{2o} \in \mathcal{H}^\infty(\mathbb{C}_+)$ are outer and that the Blaschke product $B \in \mathcal{H}^\infty(\mathbb{C}_+)$ converges in the closed right half-plane, so it is inner. Clearly the zeros of P_1 are at

$$s = \frac{-\alpha_n^4}{2}\left(\epsilon \pm \sqrt{\epsilon^2 - \frac{4}{\alpha_n^4}}\right) \qquad \text{for } n = 1, 2, \ldots \tag{7.9}$$

and those of P_2 are at

$$s = 2\alpha_n^4\left(\epsilon \pm \sqrt{\epsilon^2 + \frac{1}{\alpha_n^4}}\right) \qquad \text{for } n = 1, 2, \ldots \tag{7.10}$$

Also P_1 and P_2 each have a singularity at $-1/\epsilon$, a second order pole at $s = 0$, and poles at

$$s = \frac{-\phi_n^4}{2}\left(\epsilon \pm \sqrt{\epsilon^2 - \frac{4}{\phi_n^4}}\right) \qquad \text{for } n = 1, 2, \ldots$$

For any $\epsilon > 0$, the finite number of complex poles and zeros of P_1 alternate along the circle $|s + 1/\epsilon| = 1/\epsilon$. One branch of real poles and zeros tends to $-\infty$ and the other branch tends to $-1/\epsilon$. When $\epsilon = 0.001$, straightforward calculations show that P_1 has 13 pairs of complex poles and zeros. Examination of (7.10) shows that P_2 has all real zeros, one branch tending to $-1/\epsilon$ and the other to $+\infty$. Thus P_2 has infinitely many right half plane simple zeros, while its poles are the poles of P_1.

Note that the transfer functions $P_1(s)$ and $P_2(s)$ can be written in the form $P(s) = N(s)/D(s)$, where D is outer, and N admits an inner/outer factorization $N = M_n N_o$ where M_n is inner and N_o is outer. This notation will be used throughout the rest of this section.

7.1.2 $\mathcal{H}^\infty$ optimal control of the beam

We consider the two block $\mathcal{H}^\infty$ control problem (3.29) for the plant P (can be P_1 or P_2). Using the parametrization of Theorem 10, we know that all stabilizing controllers must be of the form

$$C = \frac{X + DQ_c}{Y - NQ_c}, \quad Q_c \in \mathcal{H}^\infty,$$

where $X, Y \in \mathcal{H}^\infty$ satisfy $NX + DY = 1$. Therefore, all admissible sensitivity functions $S = (1 + PC)^{-1}$ and complementary sensitivity functions $T = 1 - S$ in (3.29) are of the form:

$$S = 1 - N(X + DQ_c), \quad T = N(X + DQ_c), \quad Q_c \in \mathcal{H}^\infty. \tag{7.11}$$

Thus,

$$\begin{aligned} \gamma_{opt} &= \inf_{[C,P] \text{ stable}} \gamma(C) \\ &= \inf_{Q_c \in \mathcal{H}^\infty} \left\| \begin{bmatrix} W_1 \\ 0 \end{bmatrix} - \begin{bmatrix} W_1 \\ -W_2 \end{bmatrix} N(X + DQ_c) \right\|_\infty . \end{aligned} \tag{7.12}$$

Here we will consider proper stable outer weights $W_1, W_2 \in \mathcal{H}^\infty$. For P_1 or P_2 the numerator N is strictly proper. So $\gamma(C) \geq |W_1(\infty)|$ for all stabilizing controllers C. Therefore, $\gamma_{opt} \geq |W_1(\infty)|$. Similarly the properties that $D(0) = 0$ and $(1 - NX) = YD$ imply from (7.12) that $\gamma_{opt} \geq |W_2(0)|$. We will assume that

$$\gamma_{opt} > \gamma_l = \max\{|W_1(\infty)|, |W_2(0)|\}. \tag{7.13}$$

As in Chapter 5, we can reduce the problem (7.12) to a problem of the form

$$\gamma_{opt} = \inf_{Q_c \in \mathcal{H}^\infty} \left\| \begin{bmatrix} W_1^* G^{*-1} W_1 \\ W_1 W_2 G^{-1} \end{bmatrix} - \begin{bmatrix} 1 \\ 0 \end{bmatrix} GN(X + DQ_c) \right\|_\infty , \tag{7.14}$$

where G is a rational function such that $G^{-1} \in \mathcal{H}^\infty$, and

$$W_1^* W_1 + W_2^* W_2 = G^* G.$$

Again, as in Section 5.1, we construct an inner function M_w whose zeros are the right half plane poles of $W_1^* G^{*^{-1}} W_1$. Then, by defining $Q_1 := GN_o(X + DQ_c)$ it follows that

$$\gamma_{opt} \geq \gamma_1 := \inf_{Q_1 \in \mathcal{H}^\infty} \left\| \begin{bmatrix} W_o - MQ_1 \\ G_o \end{bmatrix} \right\|_\infty, \tag{7.15}$$

where

$$W_o = M_w W_1 W_1^* G^{*^{-1}} \in \mathcal{H}^\infty, \quad G_o = W_1 W_2 G^{-1}, \quad M = M_n M_w.$$

However, unlike in the case of Section 5.1, since $W_2 N_o$ is strictly proper the problem (7.15) is not equivalent to the problem (7.12). Nevertheless, in [24] and [40] it was shown that if (7.13) holds, then $\gamma_{opt} = \gamma_1$. The proof is involved, and therefore we refer the interested reader to the above cited papers. For the rest of this section we assume that (7.13) holds, so that $\gamma_{opt} = \gamma_1$ defined in (7.15). With this assumption, after finding a $Q_1^{opt} \in \mathcal{H}^\infty$ such that

$$\gamma_{opt} = \left\| \begin{bmatrix} W_o - MQ_1^{opt} \\ G_o \end{bmatrix} \right\|_\infty,$$

we can construct a candidate for the optimal controller by the formula

$$C_{opt} = N_o^{-1} D_o \frac{Q_1^{opt} G^{-1}}{1 - MQ_1^{opt} G^{-1}}. \tag{7.16}$$

The problem is now to compute γ_{opt} and Q_1^{opt}. Let us consider the following weights:

$$W_1(s) = \sqrt{b}\,\frac{as + 1/b}{as + 1} \quad W_2(s) = \sqrt{b}\,\frac{as + 1/b}{abs + 1/b}, \tag{7.17}$$

where $a > 1 \gg b > 0$. The weight W_1 penalizes the sensitivity function mainly in the low frequency range up to $(1/ab)$ rd/sec. The weight W_2 penalizes the complementary sensitivity function mainly at high frequencies above $(1/ab)$ rd/sec. This means that the frequency content of the reference signals or output disturbances are mainly below $(1/ab)$ rd/sec, while the frequency content of the measurement noise and the unmodeled dynamics are mainly above $(1/ab)$ rd/sec.

For weights with this particular structure we see that

$$\gamma_l = \sqrt{b} \quad \text{and} \quad G^{-1} = \sqrt{\frac{b}{1+b^2}} W_1^{-1} W_2^{-1}. \tag{7.18}$$

This leads to

$$G_o(s) = \sqrt{\frac{b}{1+b^2}}, \quad W_o(s) = \sqrt{\frac{b}{1+b^2}} \left(\frac{1/b - abs}{1+as} \right), \tag{7.19}$$

and

$$M_w(s) = \frac{1/b - as}{1/b + as}. \tag{7.20}$$

Since G_o is a constant, we are reduced to a one block problem:

$$\gamma_1 = \inf_{\hat{Q}_1 \in \mathcal{H}^\infty} \| \sqrt{\frac{1+b^2}{b}} W_o - M\hat{Q}_1 \|_\infty , \tag{7.21}$$

where $\hat{Q}_1 = \sqrt{\frac{1+b^2}{b}} Q_1$, and $\gamma_{opt} = \sqrt{(\gamma_1^2 + 1)b/(1+b^2)}$. We now solve this problem for the minimum phase plant $P(s) = P_1(s)$, and the non-minimum phase plant $P(s) = e^{-hs} P_2(s)$, where h is the amount of a possible time delay in the system, $h \geq 0$.

Minimum phase plant

Consider the plant $P_1 = N_1/D$, where $N_1 = P_{1o}$ is outer. In this case $M = M_w$ which is a first order inner function. It is quite easy to solve (7.21). The optimal performance is given by $\gamma_1 = \sqrt{\frac{1+b^2}{b}} |W_o(s_w)|$ where $s_w = 1/ab$ is the zero of M_w in the right half plane. Hence we have $\gamma_1 = \frac{1-b}{1+b}$, and this gives

$$\gamma_{opt} = \frac{\sqrt{2b}}{1+b}.$$

The optimal interpolant can be found as

$$Q_1^{opt} = \frac{W_o(s) - W_o(s_w)}{M_w(s)}.$$

After substitutions and simplifications we see that the optimal controller is

$$C_{opt,1} = W_1 W_2^{-1} D N_1^{-1}.$$

This leads to the following optimal sensitivity and complementary sensitivity, respectively:

$$S_{opt,1} = \frac{\sqrt{b}}{1+b} W_1^{-1}, \quad T_{opt,1} = \frac{\sqrt{b}}{1+b} W_2^{-1}.$$

In order to insure $\gamma_{opt} > \gamma_l$, b must satisfy $b < \sqrt{2} - 1$,

Non-minimum phase plant

Suppose that the plant is $P(s) = e^{-hs} P_2(s)$, where $h \geq 0$ is a possible time delay in the system, since in this case actuators and sensors are non-collocated. The general form for the inner factor of the beam transfer function is then

$$M_n(s) = e^{-hs} \prod_{n=1}^{\infty} \frac{z_n - s}{z_n + s} = e^{-hs} B(s)$$

where $h \geq 0$ and $z_n > 0$ is given in (7.10), and $z_k > z_n$, for $k > n$, with $z_n \to \infty$ as $n \to \infty$.

Since (7.21) is a one block problem we can apply the theory of Chapter 4 to solve this problem. Let us transform the problem data to unit disc via a conformal map $z = \frac{as-1}{as+1}$ where a is the weighting function parameter. With this transformation we have

$$\tilde{w}_o(z) := \sqrt{\frac{1+b^2}{b}} W_o\left(\frac{1}{a}\frac{1+z}{1-z}\right) = b_0 + b_1 z,$$

where $b_0 = (1-b^2)/2b$ and $b_1 = -(1+b^2)/2b$. From Sarason's theorem we know that γ_1 is the norm of the Sarason operator $\widetilde{w}_o(\mathbf{T})$ where $\mathbf{T}$ is the compressed shift defined on $\mathcal{H}(m)$, $m(z) = M_w(\frac{1}{a}\,\frac{1+z}{1-z})M_n(\frac{1}{a}\,\frac{1+z}{1-z})$. Note that the only essential singularity of $m(z)$ is $z = 1$, and therefore according to (4.10) the essential norm of this Sarason operator is: $|\widetilde{w}_o(1)| = b$. Also, an upper bound for the norm of $\widetilde{w}_o\mathbf{T}$ is $\|\widetilde{w}_o\|_\infty = 1/b$. The singular value/singular vector equation is in the form

$$\left(\rho^2 - \widetilde{w}_o(\mathbf{T})^*\widetilde{w}o(\mathbf{T})\right)y = 0\ , \quad y \in \mathcal{H}(m),$$

and $\rho \in (b\ ,\ 1/b)$. Defining $k(z) = 1$ and $b(z) = b_0 + b_1 z$ and applying the results of Chapter 4 we see that the above equation is equivalent to

$$\chi_\rho(z)y(z) = R_-(z)\phi_{-1} + R_+(z)\phi_0 \tag{7.22}$$

where $\chi_\rho(z) = ((b_0 z + b_1)(b_1 z + b_0) - \rho^2 z)$, $R_+(z) = b_1 b_0$, and $R_-(z) = m(z)(b_0 z + b_1)b_1 - b_1^2 m(0)$; and the unknown constants to be determined are $\phi_0 := y(0)$, and ϕ_{-1} which is the first coefficient in the power series expansion $m^* y = \sum_{k=1}^{\infty}\phi_{-k}z^{-k}$. Evaluating the right hand side of (7.22) at the roots of second order polynomial $\chi_\rho(z) = 0$, we obtain two equations in two unknowns ϕ_{-1} and ϕ_0, for $0 \neq y \in \mathcal{H}(m)$ to be a singular vector of the Sarason operator $w_o(\mathbf{T})$, corresponding to the singular value ρ. As shown in Chapter 4, after some algebra these two equations can be reduced to one equation of the form $\mathcal{R}_\rho\phi_{-1} = 0$ (note that in this example the size of $\mathcal{R}_\rho$ is 1×1). The largest value of ρ for which there exists a solution to $\mathcal{R}_\rho = 0$ determines γ_1. For the above example we can compute $\mathcal{R}_\rho$ as:

$$\mathcal{R}_\rho = \zeta_1^{-1}m(\zeta_1)(b_0\zeta_1 + b_1)b_1 - \zeta_2^{-1}m(\zeta_2)(b_0\zeta_2 + b_1)b_1 = 0$$

where

$$\zeta_1 = \beta_\rho - j\sqrt{1-\beta_\rho^2}\,, \quad \zeta_2 = \overline{\zeta}_1\,, \quad \beta_\rho = \frac{\rho^2 - b_1^2 - b_0^2}{2b_0 b_1}\,.$$

Note that $|\zeta_1| = |\zeta_2| = 1$, and $\zeta_1 = 1/\zeta_2$ for all $\rho \in (b\ ,\ 1/b)$. Hence $\mathcal{R}_\rho = 0$ is equivalent to

$$m(\zeta_2)^2\frac{(b_0 + b_1\zeta_2^{-1})}{(b_0 + b_1\zeta_2)} = 1. \tag{7.23}$$

Using (7.23), it can be shown that, in terms of the original right half plane data, γ_1 is the unique root of the equation

$$\begin{aligned}\pi &= \frac{h}{a}\omega_o + \tan^{-1}(\omega_o) + 2\tan^{-1}(b\omega_o) - \tan^{-1}(b^2\omega_o) \\ &\qquad + \sum_{n=1}^{\infty} 2\tan^{-1}(\frac{\omega_o}{az_n}),\end{aligned} \tag{7.24}$$

where

$$\omega_o = \sqrt{\frac{1/b^2 - \gamma_1^2}{\gamma_1^2 - b^2}}, \tag{7.25}$$

and $b < \gamma_1 < 1/b$. The series in (7.24) is convergent, because the Blaschke product $B(s)$ is convergent by Lemma 11. Once we find γ_1 from (7.24), the optimum controller can be computed using the procedure of Chapter 4. This is left as an exercise for the reader, see also [67].

We see that for the plant P_2, dependence of the optimal performance level on the damping coefficient ϵ enters (7.24) via the zeros z_n of the Blaschke product $B(s)$.

The optimal controllers for P_1 and P_2 have infinite number of poles and zeros, and are improper. Also, the formula for the optimal controller suggest that the double pole at $s = 0$ is cancelled by introducing a double zero in the controller at $s = 0$. This violates internal stability in the sense that $P(1 + PC_{opt})^{-1}$ does not belong to $\mathcal{H}^\infty$, because of double pole at $s = 0$. It is possible to fix this problem by either changing the weights, or by trying to find finite dimensional suboptimal controllers. It is easy to see from the above formulae that the optimal controllers for both plants P_1 and P_2 are improper. This situation occurs because the weights do not satisfy Assumption 3.2. For the beam examples given here, one needs infinite dimensional weights in order to get a proper optimal controller. However, this makes the controller computation very difficult.

On the other hand, we are interested in finite dimensional proper controllers for implementation. As discussed in Chapter 6, there are several ways to obtain such controllers. In [67], the indirect method I1 described in Chapter 6 (i.e. approximating the plant, and finding corresponding optimal controller), has been applied to this flexible beam example. The approximations of the plant were obtained by truncations of the infinite products which appear in the plant transfer functions. Convergence of the performance level of the finite dimensional proper controllers, obtained in this method, has been studied in detail, with numerical examples in [67], see also [68]. Robustness of these controllers to small time delays has also been studied [67]. See also [8] for a discussion on the same issue.

7.2 An unstable delay system

In this section we consider the following plant

$$P(s) = \frac{e^{-sh}}{s - \sigma} \tag{7.26}$$

where $h > 0$ is the time delay in the feedback loop, and $\sigma > 0$, is the right half plane pole of the system. It was shown [20] that the above plant can be taken as an abstract model of an unstable aircraft at a certain flight condition. The parameter $h\sigma$ determines the difficulty of the control problem. The control problem difficulty increases with $h\sigma$. The X-29 aircraft at its most unstable flight condition has a product of unstable pole and total time delay $h\sigma$ as large as 0.37, the other conditions being as much as a factor of 6 smaller. The meaning of "difficulty of the control" will be clear from the numerical discussion below.

In the previous chapter we have designed optimal and suboptimal controllers for the same plant, in the so called S and CS mixed sensitivity problem. Here we would like to consider the problem of robustness optimization in the gap metric, which is closely related to S and CS problem. Our purpose here is to illustrate another application of the controller formula presented in Section 6.2.3.

Let's begin by defining the robustness level $b_{P,C}$ of a closed loop system (as in Section 6.1):

$$b_{P,C}^{-1} := \left\| \begin{bmatrix} 1 \\ P \end{bmatrix} (1 + PC)^{-1} \begin{bmatrix} 1 & C \end{bmatrix} \right\|_\infty .$$

By using inner/outer factors of the plant $P = M_n N_o / M_d$ we see that

$$b_{P,C}^{-1} = \left\| \begin{bmatrix} 1 \\ N_o \end{bmatrix} (1 + PC)^{-1} \begin{bmatrix} 1 & C \end{bmatrix} \right\|_\infty .$$

Now define a spectral factor $G_P \in \mathcal{H}^\infty(\mathbb{C}_+)$ such that $G_P^{-1} \in \mathcal{H}^\infty$ and

$$G_P(s) G_P(-s) = 1 + N_o(s) N_o(-s).$$

Then, it follows that

$$b_{P,C}^{-1} = \left\| G_P \begin{bmatrix} (1 + PC)^{-1} \\ C(1 + PC)^{-1} \end{bmatrix} \right\|_\infty . \tag{7.27}$$

Now we can define $b_{opt}(P) := \sup\{b_{P,C} \; : \; [P, C] \text{ is stable}\}$ as the largest achievable robustness level in the gap metric. It is then easy to see that

$$b_{opt}^{-1} = \inf_{[P,C] \text{ is stable}} \left\| \begin{bmatrix} W_1 (1 + PC)^{-1} \\ W_2 PC(1 + PC)^{-1} \end{bmatrix} \right\|_\infty , \tag{7.28}$$

where $W_1 = G_P$ and $W_2 = G_P N_o^{-1}$. Hence the computation of $b_{opt}(P)$ amounts to solving (7.28) which is the standard S and T mixed sensitivity minimization problem (with special weights defined above), whose optimal and suboptimal solutions are summarized in Section 6.2.

Now we apply the formulae of Section 6.2, to this problem with $M_n(s) = e^{-hs}$, $M_d(s) = (s - \sigma)/(s + \sigma)$, $N_o(s) = 1/(s + \sigma)$. First we compute G_P from the spectral factorization

$$G_P(s) G_P(-s) = 1 + \frac{1}{\sigma^2 - s^2}.$$

This gives $G_P(s) = (s + \sqrt{1+\sigma^2})/(\sigma + s)$. Therefore, the weights, corresponding to the problem of robustness optimization in the gap, are

$$W_1(s) = \frac{s + \sqrt{\sigma^2+1}}{s+\sigma}, \qquad W_2(s) = s + \sqrt{\sigma^2+1}.$$

Recall that we need to construct rational functions $E_\gamma(s)$ and $F_\gamma(s)$ as defined in Section 6.2, for the search of γ_{opt} For this specific example we have

$$E_\gamma(s) = \frac{(\gamma^2-1)}{\gamma^2}\frac{(s^2-\beta_\gamma^2)}{(\sigma^2-s^2)}, \quad \beta_\gamma = \sqrt{\sigma^2 - \frac{1}{\gamma^2-1}},$$

$$F_\gamma(s) = \frac{\gamma^2}{\sqrt{\gamma^2-1}}\frac{(s-\sigma)}{(s+\sqrt{\sigma^2+1})^2}.$$

Next we determine the controller structure from Section 6.2, after simplifications we get:

$$C_{opt}(s) = \left(\frac{s-\sigma}{s+\sigma}\right)\frac{\sqrt{\gamma^2-1}(s^2-\beta_\gamma^2)L_2(s)}{(s+\sqrt{\sigma^2+1})^2L_1(s) + \frac{\gamma^2}{\sqrt{\gamma^2-1}}(s-\sigma)L_2(s)e^{-hs}}.$$

Since $n_1 = \dim(W_1) = 1$ and $\ell = \dim(M_d) = 1$ the polynomials $L_1(s)$ and $L_2(s)$ are first order, and they are determined from the interpolation conditions that the controller denominator should cancel the zeros at $s = \sigma$ and $s = \beta$. There are two more interpolation conditions to be satisfied at $s = -\sigma$ and $s = -\beta$; but these are automatically satisfied in the optimal case where $L(s) = L_2(s)/L_1(s)$ has to be all-pass. The interpolation condition at $s = \sigma$ implies that $L_1(\sigma) = 0$, which means that $L(s) = \pm(s+\sigma)/(s-\sigma)$. Thus the optimal controller has the following simplified form:

$$\pm\sqrt{\gamma_o^2-1}\frac{(s^2-\beta_{\gamma_o}^2)}{(s+\sqrt{\sigma^2+1})^2 \mp \frac{\gamma_o^2}{\sqrt{\gamma_o^2-1}}(s+\sigma)e^{-hs}} \tag{7.29}$$

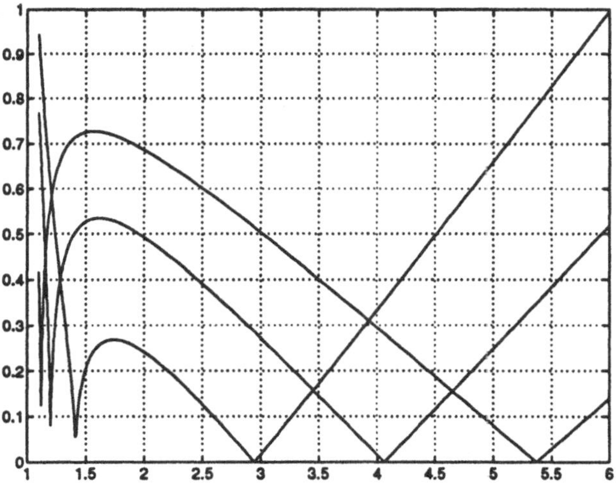

Figure 7.1: $|f(\gamma)|$ versus γ for $h = 0.1$, $\sigma = 1.0, 1.5, 2.0$.

where $\gamma_o = b_{opt}^{-1}$ is the largest value of $\gamma \geq 1$ satisfying the next equation with either the + sign or the − sign

$$f(\gamma) := (\beta_\gamma + \sqrt{\sigma^2 + 1})^2 \mp \frac{\gamma^2}{\sqrt{\gamma^2 - 1}}(\beta_\gamma + \sigma)e^{-h\beta_\gamma} = 0, \qquad (7.30)$$

with $\beta_\gamma = \sqrt{\sigma^2 - 1/(\gamma^2 - 1)}$; the optimal controller is then obtained with the appropriate sign. We see that, in order to find the optimal robustness level, all we have to do is to find the largest root γ_o of $f(\gamma)$. Then, this value of γ_o determines the optimal controller (7.29). For fixed $h = 0.1$ and typical values of $\sigma = 1.0, 1.5, 2.0$ we obtained the $f(\gamma)$ plots shown in Figure 7.1; they give $\gamma_o = 2.94, 4.06, 5.37$, respectively. For fixed $\sigma = 0.5$ and typical values of $h = 1, 1.5$ we have obtained the $f(\gamma)$ plots shown in Figure 7.2; they give $\gamma_o = 3.76, 5.05$, respectively.

The numerical illustrations indicate that as the time delay increases, and/or the magnitude of unstable pole increases the optimal robustness $b_{opt}(P)$ decreases. In other words it is difficult to control the plant P if h and/or σ is "large."

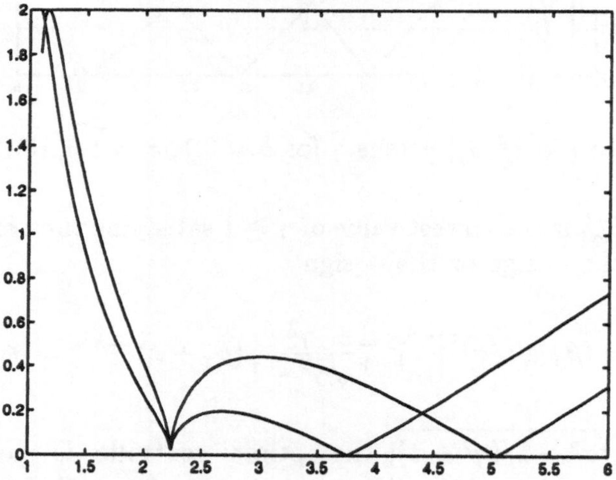

Figure 7.2: $|f(\gamma)|$ versus γ for $\sigma = 0.5$, $h = 1.0, 1.5$.

Chapter 8

$\mathcal{H}^\infty$ Control of MIMO Systems

In this chapter, we study the standard $\mathcal{H}^\infty$ control problem for MIMO distributed systems. We first will show how one can reduce the standard problem to the *four block problem* for several interesting classes of systems. We then indicate how to compute a solution to this problem.

In contrast to the SISO theory, there are still are a number of open research issues connected with multivariable systems that must be addressed before we can claim to have a satisfactory theory.

8.1 Four Block Problem

Let us recall the standard $\mathcal{H}^\infty$ problem from Chapter 3. Consider the feedback system in Figure 3.2, where G represents the generalized plant to be controlled (containing the plant and the weights modelling the disturbances). The signal v represents the exogenous inputs; u is the control input; z is the output to be controlled; and y is the measured output. The standard $\mathcal{H}^\infty$ control problem is to minimize the $\mathcal{H}^\infty$ norm of the transfer function T_{zv}, from v to z, over all internally stabilizing controllers K:

$$\gamma_{opt} := \inf_K \|T_{zv}\|_\infty.$$

Under very mild hypotheses, for a linear time invariant G, one can show that the above problem is equivalent to the following (see e.g. [39])

$$\gamma_{opt} = \inf_{Q\in\mathcal{H}^\infty_{n\times m}} \|T_1 - T_2QT_3\|_\infty \tag{8.1}$$

where $T_1 \in \mathcal{H}^\infty_{n_2\times n_3}$, $T_2 \in \mathcal{H}^\infty_{n_2\times n}$, $T_3 \in \mathcal{H}^\infty_{m\times n_3}$ are determined from the weights and Youla parametrization of all stabilizing controllers, [91, 116] and Q is the free parameter. Without loss of generality, we will assume that $n_2 \geq n$ and $n_3 \geq m$, i.e. T_2 is "tall" and T_3 is "fat."

Now, T_2 admits an inner/outer factorization, [94], in the form $T_2 = T_{2i}T_{2o}$ where $T_{2i} \in \mathcal{H}^\infty_{n_2\times n}$ is inner and $T_{2o} \in \mathcal{H}^\infty_{n\times n}$ is outer. Similarly we can find a co-inner/co-outer factorization of $T_3 = T_{3o}T_{3i}$, where $T_{3o} \in \mathcal{H}^\infty_{m\times m}$ is co-outer, and $T_{3i} \in \mathcal{H}^\infty_{m\times n_3}$ is co-inner (i.e. its transpose is inner). There exists an inner (resp. co-inner) matrix $T_{2i\perp}$ (resp. $T_{3i\perp}$) such that $[T_{2i} \quad T_{2i\perp}]$ (resp. $\begin{bmatrix} T_{3i} \\ T_{3i\perp}\end{bmatrix}$) is square inner. We now have

$$\gamma_{opt} = \inf_{Q\in\mathcal{H}^\infty_{n\times m}} \left\| [T_{2i} \quad T_{2i\perp}]^* \, (T_1 - T_2QT_3) \begin{bmatrix} T_{3i} \\ T_{3i\perp}\end{bmatrix}^* \right\|_\infty$$

which is equivalent to

$$\gamma_{opt} = \inf_{Q_1\in\mathcal{H}^\infty_{n\times m}} \left\| \begin{bmatrix} T_{2i}^*T_1T_{3i}^* - Q_1 & T_{2i}^*T_1T_{3i\perp}^* \\ T_{2i\perp}^*T_1T_{3i}^* & T_{2i\perp}^*T_1T_{3i\perp}^* \end{bmatrix} \right\|_\infty \tag{8.2}$$

where $Q_1 = T_{2o}QT_{3o}$. Using spectral factorizations of the entries of the above matrix one can find inner matrices $M \in \mathcal{H}^\infty_{n\times n}$ and $M_1 \in \mathcal{H}^\infty_{\kappa\times\kappa}$, $\kappa := n_2 - n$ such that

$$\begin{aligned} W &:= MT_{2i}^*T_1T_{3i}^* \in \mathcal{H}^\infty_{n\times m}\,, \quad F := MT_{2i}^*T_1T_{3i\perp}^* \in \mathcal{H}^\infty_{n\times\ell}\,, \\ G &:= M_1T_{2i\perp}^*T_1T_{3i}^* \in \mathcal{H}^\infty_{\kappa\times m}\,, \quad J := M_1T_{2i\perp}^*T_1T_{3i\perp}^* \in \mathcal{H}^\infty_{\kappa\times\ell} \end{aligned}$$

where $\ell := n_3 - m$. Then, we multiply the matrix on the right hand side of (8.2) by $\begin{bmatrix} M & 0 \\ 0 & M_1 \end{bmatrix}$ from the left and obtain

$$\gamma_{opt} = \inf_{Q_1 \in \mathcal{H}^{\infty}_{n \times m}} \left\| \begin{bmatrix} W - MQ_1 & F \\ G & J \end{bmatrix} \right\|_{\infty}. \tag{8.3}$$

The problems (8.2) and (8.3) are called the four block problems.

8.2 Computation of γ_{opt}

In this section we will give the statement of commutant lifting theorem. Then, by using this result we will obtain three different operators, namely the four block operator, Young's operator, and Hankel operator, whose norms determine γ_{opt}.

8.2.1 Commutant Lifting Theorem

In this section, we formulate and prove the commutant lifting theorem which forms the theoretical basis of this monograph. This theorem is proven in a very general context and is directed to the reader who is interested in the functional analytical underpinnings of our approach. Throughout this section, $\mathcal{H}$ will denote a complex separable Hilbert space. By "operator" we shall always mean "bounded linear operator," unless explicitly stated otherwise. By using the commutant lifting theorem we show how the solution to the four block problem may be reduced to computing the norm of a certain operator derived from this theorem.

We first begin with a key result due to Sz.-Nagy, see [94] Chapter I. Let $\mathbf{T}' : \mathcal{H}' \to \mathcal{H}'$ be a contraction, i.e. an operator such that $\|\mathbf{T}'\| \leq 1$. Then one can prove that there exists an isometry $\mathbf{U}'$ on a Hilbert space $\mathcal{K}'$ such that

$$\mathcal{K}' = \bigvee_{n=0}^{\infty} \mathbf{U}'^n \mathcal{H}'$$

and $\mathbf{P}_{\mathcal{H}'}\mathbf{U}' = \mathbf{T}'\mathbf{P}_{\mathcal{H}'}$, where $\mathbf{P}_{\mathcal{H}'} : \mathcal{K}' \to \mathcal{H}'$ denotes orthogonal projection; and $\mathbf{U}'$ is called the *minimal isometric dilation of* $\mathbf{T}'$. To construct it, take the positive square root $\mathbf{D}_{\mathbf{T}'} = (\mathbf{I} - \mathbf{T}'^*\mathbf{T}')^{1/2}$ of $(\mathbf{I} - \mathbf{T}'^*\mathbf{T}')$, which is positive definite since $\mathbf{T}'$ is a contraction. Define

$$\mathcal{K}' = \mathcal{H}' \oplus \mathcal{H}^2(\mathsf{D}, \mathcal{D}_{\mathbf{T}'})$$

where $\mathcal{H}^2(\mathsf{D}, \mathcal{D}_{\mathbf{T}'})$ denotes the Hardy space $\mathcal{H}^2$ on D formed by analytic functions in D with values in $\mathcal{D}_{\mathbf{T}'}$, closure of the range of $\mathbf{D}_{\mathbf{T}'}$. Recall that

$$h = \sum_{n=0}^{\infty} z^n h_n \in \mathcal{H}^2(\mathsf{D}, \mathcal{D}_{\mathbf{T}'})$$

means that $h_n \in \mathcal{D}_{\mathbf{T}'}$ and that

$$\|h\|^2 := \sum_{n=0}^{\infty} \|h_n\|^2 < \infty .$$

On $\mathcal{K}'$ define

$$\mathbf{U}'(h' \oplus d) = \mathbf{T}'h' \oplus (\mathbf{D}_{\mathbf{T}'}h + zd(z))$$

and identify $\mathcal{H}'$ with the subspace $\mathcal{H}' \oplus \{0\}$ of $\mathcal{K}'$. One can easily check that $\mathbf{U}'$ is the minimal isometric dilation of $\mathbf{T}'$. For instance

$$\begin{aligned}
\|\mathbf{U}'(h' \oplus d)\|^2 &= \|\mathbf{T}'h\|^2 + \|\mathbf{D}_{\mathbf{T}'}h + zd(z)\|^2 \\
&= \|\mathbf{T}'h\|^2 + \|\mathbf{D}_{\mathbf{T}'}h\|^2 + \|d_0\|^2 + \|d_1\|^2 + \cdots \\
&= (\mathbf{T}'^*\mathbf{T}'h, h) + (\mathbf{D}^2_{\mathbf{T}'}h, h) + \|d(\cdot)\|^2 \\
&= \|h\|^2 + \|d(\cdot)\|^2 \quad \forall\, h \oplus d \in \mathcal{K}'
\end{aligned}$$

so $\mathbf{U}'$ is isometric. Then the *commutant lifting theorem* may be stated as follows (see [93], [94] Chapter II, and also [28]).

Theorem 25 (Commutant Lifting Theorem) *Let $\mathcal{H}$ and $\mathcal{H}'$ denote (complex separable) Hilbert spaces with $\mathbf{T} : \mathcal{H} \to \mathcal{H}$, $\mathbf{T}' : \mathcal{H}' \to \mathcal{H}'$ contractions. Let $\mathbf{A} : \mathcal{H} \to \mathcal{H}'$ be a contraction intertwining $\mathbf{T}$ and*

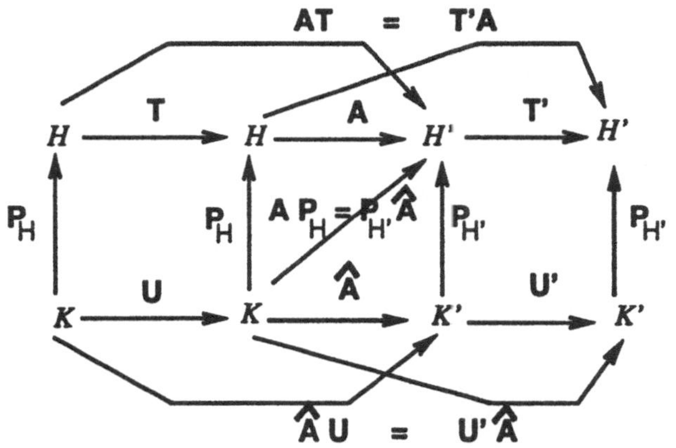

Figure 8.1: Operators of the Commutant Lifting Theorem

$\mathbf{T}'$, *i.e.* $\mathbf{AT} = \mathbf{T}'\mathbf{A}$. *Let* $\mathbf{U} : \mathcal{K} \to \mathcal{K}$ *and* $\mathbf{U}' : \mathcal{K}' \to \mathcal{K}'$ *denote the minimal isometric dilations of* $\mathbf{T}$ *and* $\mathbf{T}'$ *respectively. Then there exists a contraction* $\widehat{\mathbf{A}} : \mathcal{K} \to \mathcal{K}'$ *such that* $\mathbf{U}'\widehat{\mathbf{A}} = \widehat{\mathbf{A}}\mathbf{U}$ *and* $\mathbf{P}_{\mathcal{H}'}\widehat{\mathbf{A}} = \mathbf{AP}_{\mathcal{H}}$ *where* $\mathbf{P}_{\mathcal{H}'} : \mathcal{K}' \to \mathcal{H}'$ *and* $\mathbf{P}_{\mathcal{H}} : \mathcal{K} \to \mathcal{H}$ *denote orthogonal projections.*

Remarks

(i) $\widehat{\mathbf{A}}$ is called an *intertwining dilation* of $\mathbf{A}$. All the spaces and operators defined in the above theorem and their relations are illustrated in Figure 8.1.

(ii) For the commutant lifting theorem we do not need the minimal isometric dilation, but any isometric dilation of $\mathbf{T}$ satisfying $\mathbf{T}^* = \mathbf{U}^*|\mathcal{H}$. Moreover, in the proof we may take without loss of generality $\mathcal{K} = \mathcal{H}$, $\mathbf{U} = \mathbf{T}$ and replace the original $\mathbf{A}$ by $\mathbf{AP}_{\mathcal{H}}$. Indeed define $\mathbf{A}_\nu := \mathbf{AP}_{\mathcal{H}}$, then we have $\|\mathbf{A}_\nu\| = \|\mathbf{A}\|$, and

$$\mathbf{A}_\nu\mathbf{U} = \mathbf{AP}_{\mathcal{H}}\mathbf{U} = \mathbf{ATP}_{\mathcal{H}} = \mathbf{T}'\mathbf{AP}_{\mathcal{H}} = \mathbf{T}'\mathbf{A}_\nu\ .$$

Therefore, $\mathbf{A} := \mathbf{A}_\nu$ satisfies the requirements of the theorem in this special case.

The power of the commutant lifting theorem is that it allows one to derive most of the classical interpolation results involving bounded analytic functions in a unified, elegant manner even when the functions are operator-valued.

In the next section we will show how the solution of the four block problem may be reduced to computing the norm of the four block operator via the commutant lifting theorem. We now give a proof of the commutant lifting theorem.

A Proof of the Commutant Lifting Theorem

As above, without loss of generality take $\mathcal{K} = \mathcal{H}$, $\mathbf{T} = \mathbf{U}$, see (ii).

For $h \in \mathcal{H} = \mathcal{K}$ we look for an $\widehat{\mathbf{A}}h$ of the form

$$\widehat{\mathbf{A}}h = \mathbf{A}h \oplus \sum_{n=0}^{\infty} z^n B_n h,$$

for some operators $\{B_n \ : \ n = 0, 1, \ldots\}$ from $\mathcal{H}$ into $\mathcal{D}_{\mathbf{T}'}$. In what follows $\|\cdot\|$ denotes the norm of the various vectors, functions and operators. The meaning should be clear from the context. By the above

$$\|h\|^2 \geq \|\widehat{\mathbf{A}}h\|^2 = \|\mathbf{A}h\|^2 + \sum_{n=0}^{\infty} \|B_n h\|^2$$

which implies that

$$\langle (\mathbf{I} - \mathbf{A}^*\mathbf{A})h, h \rangle \geq \sum_{n=0}^{\infty} \|B_n h\|^2$$

Let $\mathbf{D_A}$ be the square root of $(\mathbf{I} - \mathbf{A}^*\mathbf{A})$. Then, the above inequality means that

$$B_n h = C_n \mathbf{D_A} h$$

for some $\{C_n : n = 0, 1, \ldots\}$, $C_n : \mathcal{D}_{\mathbf{A}} \to \mathcal{D}_{\mathbf{T}'}$, where $\mathcal{D}_{\mathbf{A}}$ is the closure of the range of $\mathbf{D}_{\mathbf{A}}$. In summary, we must have

$$\widehat{\mathbf{A}}h = \mathbf{A}h \oplus \sum_{n=0}^{\infty} z^n C_n \mathbf{D}_{\mathbf{A}} h \quad \forall\, h \in \mathcal{H}\,, \tag{8.4}$$

$$\|d\|^2 \geq \sum_{n=0}^{\infty} \|C_n d\|^2 \quad \forall\, d \in \mathcal{D}_{\mathbf{A}}. \tag{8.5}$$

Since

$$\begin{aligned} \widehat{\mathbf{A}}\mathbf{T}h &= \mathbf{A}\mathbf{T}h \oplus \sum_{n=0}^{\infty} z^n C_n \mathbf{D}_{\mathbf{A}}\mathbf{T}h \qquad \text{and} \\ \mathbf{U}'\widehat{\mathbf{A}}h &= \mathbf{T}'\mathbf{A}h \oplus \left(\mathbf{D}_{\mathbf{T}'}\mathbf{A}h + z\sum_{n=0}^{\infty} z^n C_n \mathbf{D}_{\mathbf{A}}h\right)\,, \end{aligned}$$

we see that for the condition $\widehat{\mathbf{A}}\mathbf{T} = \mathbf{U}'\widehat{\mathbf{A}}$ to hold it is necessary and sufficient that

$$C_0 \mathbf{D}_{\mathbf{A}}\mathbf{T} = \mathbf{D}_{\mathbf{T}'}\mathbf{A} \tag{8.6}$$

$$C_{n+1}\mathbf{D}_{\mathbf{A}}\mathbf{T} = C_n \mathbf{D}_{\mathbf{A}} \qquad n = 0, 1, \ldots \tag{8.7}$$

Thus the construction of $\widehat{\mathbf{A}}$ is reduced to finding $C(z) = \sum_{n=0}^{\infty} z^n C_n$ satisfying (8.5), (8.6) and (8.7). Now define the operator ω as follows

$$\omega \mathbf{D}_{\mathbf{A}}\mathbf{T}h := \mathbf{D}_{\mathbf{A}}h \oplus \mathbf{D}_{\mathbf{T}'}\mathbf{A}h\,, \quad \forall\, h \in \mathcal{H}$$

Then, the following equalities hold:

$$\begin{aligned} \|\omega \mathbf{D}_{\mathbf{A}}\mathbf{T}h\|^2 &= \|\mathbf{D}_{\mathbf{A}}h\|^2 + \|\mathbf{D}_{\mathbf{T}'}\mathbf{A}h\|^2 \\ &= \|h\|^2 - \|\mathbf{A}h\|^2 + \|\mathbf{A}h\|^2 - \|\mathbf{T}'\mathbf{A}h\|^2 \\ &= \|h\|^2 - \|\mathbf{A}\mathbf{T}h\|^2 \\ &= \|\mathbf{T}h\|^2 - \|\mathbf{A}\mathbf{T}h\|^2 \\ &= \|\mathbf{D}_{\mathbf{A}}\mathbf{T}h\|^2\,. \end{aligned}$$

Hence ω is an isometry.

Now let $\mathcal{F}$ be the closure of $\{\mathbf{D}_{\mathbf{A}}\mathbf{T}h : h \in \mathcal{H}\} \subset \mathcal{D}_{\mathbf{A}}$, and $\mathcal{F}' = \omega\mathcal{F} \subset \mathcal{D}_{\mathbf{A}} \oplus \mathcal{D}_{\mathbf{T}'}$. Let $W(z)$ be an analytic function in the unit disc, that is

$$W(z) = W_0 + zW_1 + z^2 W_2 + \cdots \quad \text{for} \quad z \in \mathsf{D}\,,$$

with values operators from $\mathcal{D}_{\mathbf{A}}$ into $\mathcal{D}_{\mathbf{A}} \oplus \mathcal{D}_{\mathbf{T}'}$ such that $\|W\|_\infty = \sup_{|z|=1} \|W(z)\| \leq 1$, and $W(z)f = \omega f$ for all $z \in \mathbf{D}$ and $f \in \mathcal{F}$. Note that this last condition means that $W_0\mathbf{D}_{\mathbf{A}}\mathbf{T} = \omega\mathbf{D}_{\mathbf{A}}\mathbf{T}$, $W_n\mathbf{D}_{\mathbf{A}}\mathbf{T} = 0$ for $n = 1, 2 \ldots$. For instance the constant function $W(z) = \omega\mathbf{P}_{\mathcal{F}}$ satisfies these conditions. Also define the operators $\mathbf{\Pi}_{\mathbf{A}}$ and $\mathbf{\Pi}_{\mathbf{T}'}$ on $\mathcal{F}'$ by

$$\begin{aligned} \mathbf{\Pi}_{\mathbf{A}}(d \oplus h') &= d \\ \mathbf{\Pi}_{\mathbf{T}'}(d \oplus h') &= h' \end{aligned}$$

for $d \in \mathcal{D}_{\mathbf{A}}$ and $h' \in \mathcal{D}_{\mathbf{T}'}$. Now introduce $D(z)$ defined by

$$(\mathbf{I} - z\mathbf{\Pi}_{\mathbf{A}}W(z))D(z) = I$$

more explicitly

$$(I - z\mathbf{\Pi}_{\mathbf{A}}W_0 - z^2\mathbf{\Pi}_{\mathbf{A}}W_1 - \cdots)(D_0 + zD_1 + \cdots) = I .$$

The above equation means that $D_0 = I$ and

$$D_{n+1} - \mathbf{\Pi}_{\mathbf{A}}W_0D_n - \mathbf{\Pi}_{\mathbf{A}}W_1D_{n-1} - \cdots \mathbf{\Pi}_{\mathbf{A}}W_nD_0 = 0 \tag{8.8}$$

By induction it is easy to show that

$$D_0 = I \quad \text{and} \quad D_{n+1}\mathbf{D}_{\mathbf{A}}Th = D_n\mathbf{D}_{\mathbf{A}}h \qquad n = 0, 1, 2, \ldots \tag{8.9}$$

In terms of $D(z)$, whose solution is given by (8.9), we define $C(z)$ as

$$C(z) := \mathbf{\Pi}_{\mathbf{T}'}W(z)D(z) . \tag{8.10}$$

Using the above definitions, (8.8) and (8.9) it is easy to check that the conditions (8.6), (8.7) are satisfied. Hence to complete the proof it is left to show that $C(z)$, defined by (8.10), is a contraction, that is, it satisfies (8.5). To this aim, take any $d := d_0 + zd_1 + \cdots + z^N d_N \in \mathcal{H}^2(\mathcal{D}_{\mathbf{A}})$ with $d_n = D_n d_0$ for $n = 0, 1, \ldots, N$. Then, by the definition of W

$$\sum_{n=0}^{N} \|d_n\|^2 \geq \sum_{n=0}^{\infty} \|(Wd)_n\|^2$$

where $(Wd)_n = W_n d_0 + \cdots + W_0 d_n = (WD)_n d_0$ for $n \leq N$. But

$$\sum_{n=0}^{N} (\|\Pi_{\mathbf{T}'}(WD)_n d_0\|^2 + \|\Pi_{\mathbf{A}}(WD)_n d_0\|^2) = \sum_{n=0}^{N} \|(WD)_n d_0\|^2$$

$$= \sum_{n=0}^{N} \|(Wd)_n\|^2 \leq \sum_{n=0}^{N} \|D_n d_0\|^2$$

so

$$\sum_{n=0}^{N} \|C_n d_0\|^2 \leq \sum_{n=0}^{N} \|D_n d_0\|^2 - \sum_{n=0}^{N} \|\Pi_{\mathbf{A}}(WD)_n d_0\|^2$$

and, by the identity $\Pi_{\mathbf{A}}(WD)_n d_0 = D_{n+1} d_0$, (see (8.8)),

$$\sum_{n=0}^{N} \|C_n d_0\|^2 \leq \sum_{n=0}^{N} \|D_n d_0\|^2 - \sum_{n=0}^{N} \|D_{n+1} d_0\|^2 .$$

Therefore,

$$\sum_{n=0}^{N} \|C_n d_0\|^2 \leq \|d_0\|^2 - \|D_{N+1} d_0\|^2 \leq \|d_0\|^2$$

which concludes the proof.

Remark. A perusal of the preceding proof shows that if $W(z)$ is an analytic function in the unit disc, that is

$$W(z) = W_0 + zW_1 + z^2 W_2 + \cdots \quad \text{for} \quad z \in \mathbf{D} ,$$

with values operators from $\mathcal{D}_{\mathbf{A}}$ into $\mathcal{D}_{\mathbf{A}} \oplus \mathcal{D}_{\mathbf{T}'}$ such that $\|W\|_\infty = \sup_{|z|=1} \|W(z)\| \leq 1$, and $W(z)f = \omega f$ for all $z \in \mathbf{D}$ and $f \in \mathcal{F}$, then

$$\widehat{\mathbf{A}} = \begin{bmatrix} \mathbf{A} \\ \Pi_{\mathbf{T}'} W(z)(I - z\Pi_{\mathbf{A}} W(z))^{-1} \mathbf{D}_{\mathbf{A}} \end{bmatrix}$$

from $\mathcal{H}$ into $\mathcal{K}' = \mathcal{H} \oplus \mathcal{H}^2(\mathcal{D}_{\mathbf{T}'})$ is a contraction intertwining $\mathbf{U}'$ with $\mathbf{T}$ such that $\mathbf{P}_{\mathcal{H}} \widehat{\mathbf{A}} = \mathbf{A}$. It can be shown that any such operator $\widehat{\mathbf{A}}$ can be obtained in this way with a unique $W(z)$ of the type described above (see [28], Chapter XIV; actually the proof given above also follows [28], Chapter XIV).

8.2.2 Four Block Operator

Recall the four block problem defined by (8.3), where the (1,1) block is $n \times m$, the (1,2) block is $n \times \ell$, the (2,1) block is $\kappa \times m$, and the (2,2) block is $\kappa \times \ell$. Note that original problem data has the following dimensions: T_1 is $n_2 \times n_3$, T_2 is $n_2 \times n$, T_3 is $m \times n_3$, and the free parameter Q is $n \times m$. We have assumed that $\ell = n_3 - m \geq 0$, and $\kappa = n_2 - n \geq 0$. In this section, we assume that all the matrices are defined in terms of the z variable on the unit disc (so a continuous time problem data has already been transformed to the unit disc via a conformal map). Therefore, all Hardy (resp., Lebesgue) spaces are defined on the unit disc (resp., unit circle).

We denote the canonical unilateral shift (defined by multiplication by z) on $\mathcal{H}^2_m$ by $\mathbf{S}$, the bilateral shift on $\mathcal{L}^2_\ell$ by $\mathbf{U}$ and the bilateral shift on $\mathcal{L}^2_\kappa$ by $\mathbf{U}'$.

Recall that the four block $\mathcal{H}^\infty$ problem amounts to finding

$$\gamma_{opt} := \inf \left\{ \left\| \begin{bmatrix} W - MQ & F \\ G & J \end{bmatrix} \right\|_\infty \ : \ Q \in \mathcal{H}^\infty_{n \times m} \right\}.$$

We would like to remind the reader that for an $n_2 \times n_3$ matrix of the form $\begin{bmatrix} A & B \\ C & D \end{bmatrix}$, where A, B, C, D having appropriate sizes with entries in $\mathcal{L}^\infty$ the ∞-norm is defined as

$$\left\| \begin{bmatrix} A & B \\ C & D \end{bmatrix} \right\|_\infty = \text{ess sup} \left\{ \left\| \begin{bmatrix} A(\zeta) & B(\zeta) \\ C(\zeta) & D(\zeta) \end{bmatrix} \right\| : \zeta \in \mathbf{T} \right\}.$$

For the norm on the right hand side the $n_2 \times n_3$ matrix is taken as a linear operator from $\mathbb{C}^{n_3}$ to $\mathbb{C}^{n_2}$ for each fixed ζ in $\mathbf{T}$. Note that if $F = G = J = 0$ then this problem reduces to the *one block problem*. For $F = J = 0$ we have the *two block problem*.

To the $n \times n$ inner matrix M, we associate the spaces $\mathcal{H}(M) := \mathcal{H}^2_n \ominus M\mathcal{H}^2_n$ and $\mathcal{L}(M) := \mathcal{L}^2_n \ominus M\mathcal{H}^2_n$. Let $\mathbf{P}_{\mathcal{H}(M)} : \mathcal{H}^2_n \to \mathcal{H}(M)$, and $\mathbf{P}_{\mathcal{L}(M)} : \mathcal{L}^2_n \to \mathcal{L}(M)$, be the orthogonal projections.

We now define the four block operator (see [22, 34, 81]): $\mathbf{A} : \mathcal{H}^2_m \oplus \mathcal{L}^2_\ell \to \mathcal{L}(M) \oplus \mathcal{L}^2_\kappa$ as follows

$$\mathbf{A} := \begin{bmatrix} \mathbf{P}_{\mathcal{H}(M)} W(\mathbf{S}) & \mathbf{P}_{\mathcal{L}(M)} F(\mathbf{U}) \\ G(\mathbf{S}) & J(\mathbf{U}) \end{bmatrix} = \begin{bmatrix} \mathbf{P}_{\mathcal{L}(M)} & 0 \\ 0 & \mathbf{I} \end{bmatrix} \begin{bmatrix} W(\mathbf{S}) & F(\mathbf{U}) \\ G(\mathbf{S}) & J(\mathbf{U}) \end{bmatrix}.$$

Above, as in the sequel we denote by $G(\mathbf{S})$ the multiplication operator by $G(\zeta)$ from $\mathcal{H}^2_\kappa$ into $\mathcal{L}^2_m$ while $F(\mathbf{U})$ denotes the multiplication operator by $F(\zeta)$ from $\mathcal{L}^2_\ell$ into $\mathcal{L}^2_n$, similarly for $W(\mathbf{S})$ and $J(\mathbf{U})$, and later for $W_1(\mathbf{S})$, $F_1(\mathbf{U})$, etc. In this section, by a slight abuse of notation, ζ will denote a complex variable as well as an element of $\mathbf{T}$. The context will make the meaning clear. We shall now employ the commutant lifting theorem to show that γ_{opt} is equal to $\|\mathbf{A}\|$.

Theorem 26 *Notation as above. Then*

$$\|\mathbf{A}\| = \gamma_{opt}.$$

Proof. First since for an arbitrary $Q \in \mathcal{H}^\infty_{n\times m}$ and $u \in \mathcal{H}^2_m$, we have

$$\mathbf{P}_{\mathcal{L}(M)}(MQu) = 0$$

from (8.3) and the definition of $\mathbf{A}$, this implies that $\gamma_{opt} \geq \|\mathbf{A}\|$. Thus in order to prove the theorem, we need to show the existence of a matrix $Q_{opt} \in \mathcal{H}^\infty_{n\times m}$ such that

$$\left\| \begin{bmatrix} W - MQ_{opt} & F \\ G & J \end{bmatrix} \right\|_\infty = \|\mathbf{A}\|.$$

Let

$$\mathbf{U}_1 := \begin{bmatrix} \mathbf{S} & \mathbf{0} \\ \mathbf{0} & \mathbf{U} \end{bmatrix} \quad \text{and} \quad \mathbf{T}_2 := \begin{bmatrix} \mathbf{P}_{\mathcal{L}(M)} \mathbf{U} | \mathcal{L}(M) & \mathbf{0} \\ \mathbf{0} & \mathbf{U}' \end{bmatrix}.$$

Note that $\mathbf{U}_1 : \mathcal{H}^2_m \oplus \mathcal{L}^2_\ell \to \mathcal{H}^2_m \oplus \mathcal{L}^2_\ell$ and $\mathbf{T}_2 : \mathcal{L}(M) \oplus \mathcal{L}^2_\kappa \to \mathcal{L}(M) \oplus \mathcal{L}^2_\kappa$. Now it is easy to see that we have $\mathbf{A}\mathbf{U}_1 = \mathbf{T}_2\mathbf{A}$. Applying the commutant lifting theorem , we can then deduce the existence of an intertwining dilation $\mathbf{B} : \mathcal{H}^2_m \oplus \mathcal{L}^2_\ell \to \mathcal{L}^2_m \oplus \mathcal{L}^2_\kappa$, i.e., an operator with the properties that

$$\begin{bmatrix} \mathbf{P}_{\mathcal{L}(M)} & \mathbf{0} \\ \mathbf{0} & \mathbf{I} \end{bmatrix} \mathbf{B} = \mathbf{A} \tag{8.11}$$

and

$$\mathbf{B} \begin{bmatrix} \mathbf{S} & \mathbf{0} \\ \mathbf{0} & \mathbf{U} \end{bmatrix} = \begin{bmatrix} \mathbf{U} & \mathbf{0} \\ \mathbf{0} & \mathbf{U}' \end{bmatrix} \mathbf{B}. \tag{8.12}$$

Moreover $\mathbf{B}$ can be chosen such that

$$\|\mathbf{B}\| = \|\mathbf{A}\|. \tag{8.13}$$

But notice that from (8.12),

$$\mathbf{B} = \begin{bmatrix} W_1(\mathbf{S}) & F_1(\mathbf{U}) \\ G_1(\mathbf{S}) & J_1(\mathbf{U}) \end{bmatrix}$$

where W_1 is an $\mathcal{H}^\infty$ matrix-valued function of size $n \times m$ and and G_1, F_1, J_1 are $\mathcal{L}^\infty$ matrix-valued functions of size $\kappa \times m$, $n \times \ell$ and $\kappa \times \ell$, respectively. Moreover,

$$\left\| \begin{bmatrix} W_1 & F_1 \\ G_1 & J_1 \end{bmatrix} \right\|_\infty = \|\mathbf{B}\|.$$

But, from (8.11), we have that $G = G_1$, $J = J_1$, and

$$\mathbf{P}_{\mathcal{L}(M)} F(\mathbf{U}) = \mathbf{P}_{\mathcal{L}(M)} F_1(\mathbf{U}) \tag{8.14}$$

$$\mathbf{P}_{\mathcal{L}(M)} W(\mathbf{S}) = \mathbf{P}_{\mathcal{L}(M)} W_1(\mathbf{S}) \tag{8.15}$$

The relation (8.15) implies that $W_1 = W - MQ_{opt}$ for some $Q_{opt} \in \mathcal{H}^\infty_{n \times m}$; (the proof for the scalar case was given in the remark (iv) in

Section 2.7). The proof for the matrix case is rather similar (see also Chapter IX, Section 2 of [28]). Thus in order to complete the proof of the theorem we must show that (8.14) implies that $F = F_1$. But certainly, from (8.14) we can immediately infer that

$$\Delta_F(\mathbf{U})v \in M\mathcal{H}^2_n, \quad \forall v \in \mathcal{L}^2_\ell,$$

where $\Delta_F := F - F_1$. This means that

$$(M^*\Delta_F)(\mathbf{U})v \in \mathcal{H}^2_n, \quad \forall v \in \mathcal{L}^2_\ell.$$

Thus, since $\zeta^k v \in \mathcal{L}^2_\ell$ for all $k = 0, \pm 1, \pm 2, \ldots$ we also have

$$\zeta^k M^*\Delta_F(\mathbf{U})v \in \mathcal{H}^2_n \quad \forall\, k = 0, \pm 1, \pm 2, \ldots$$

This forces

$$(M^*\Delta_F)(\mathbf{U})v = 0, \quad \forall v \in \mathcal{L}^2_\ell \ominus \mathcal{H}^2_\ell.$$

Hence $\Delta_F v = 0$ for all v, this implies $\Delta_F = 0$ which completes the proof. □

In [81] there is a detailed discussion on how to find the singular values of the four block operator **A**. This requires certain assumptions on the structure of the problem data, e.g. W, G, F, J are rational, and certain commutativity assumptions. For details see [58] and [81].

8.2.3 Young's Operator

A key difficulty involved in reducing the standard problem (8.1) to the four block problem (8.2), or (8.3), is the various kinds of factorizations that must be performed. In fact, one of the major advantages involved in the recent state space methods is that these factorizations may be avoided. Of course as mentioned before, one of the disadvantages of these state space methods is that their practical applicability to distributed systems seems to be very difficult. On an infinite dimensional state space one gets infinite dimensional, i.e., operator-valued Riccati equations.

In this section, we would like to describe an operator first defined by N. Young, [118] which in principle can avoid a number of the problems with such factorizations (especially in the multivariable distributed case), as well as allow the utilization of our frequency domain "skew Toeplitz" methods to distributed systems. First of all, recall that via the Youla parametrization, the standard problem may be formulated as in (8.1), where T_1, T_2, T_3, Q are matrix-valued $\mathcal{H}^\infty$ functions (whose sizes are defined above). More precisely, we take $T_1 \in \mathcal{H}^\infty_{n_2 \times n_3}$, $T_2 \in \mathcal{H}^\infty_{n_2 \times n}$, $T_3 \in \mathcal{H}^\infty_{m \times n_3}$, and the parameter $Q \in \mathcal{H}^\infty_{n \times m}$. For simplicity, we assume that T_2 is inner, and that T_3 is co-inner, i.e. inner outer factorization of T_2 and co-inner co-outer factorization of T_3 are already made and the outer factors are absorbed into Q.

Define

$$\begin{aligned}
\mathcal{K}^2_m &:= \mathcal{L}^2_m \ominus \mathcal{H}^2_m \,, \\
\mathcal{H}_{T_3} &:= \mathcal{L}^2_{n_3} \ominus T_3^* \mathcal{K}^2_m \,, \\
\mathcal{H}_{T_2} &:= \mathcal{L}^2_{n_2} \ominus T_2 \mathcal{H}^2_n \,,
\end{aligned}$$

and the operator $\mathbf{\Lambda}_{T_1} : \mathcal{H}_{T_3} \to \mathcal{H}_{T_2}$ by

$$\mathbf{\Lambda}_{T_1} f := \mathbf{P}_{\mathcal{H}_{T_2}} T_1(\mathbf{U}) f$$

where $\mathbf{P}_{\mathcal{H}_{T_2}}$ denotes orthogonal projection on $\mathcal{H}_{T_2}$. The operator $\mathbf{\Lambda}_{T_1}$ is known as *Young's operator* associated with the present standard problem.

Theorem 27 *For T_1, T_2 and T_3 given as above we have*

$$\inf_{Q \in \mathcal{H}^\infty_{n \times m}} \|T_1 - T_2 Q T_3\|_\infty = \|\mathbf{\Lambda}_{T_1}\|.$$

Proof. The proof is very similar to that of Theorem 26 so we just outline the proof. First since $\mathbf{\Lambda}_{T_2 Q T_3} = \mathbf{0}$ for any $Q \in \mathcal{H}^\infty_{n \times m}$, we see that

$$\begin{aligned}\|\mathbf{\Lambda}_{T_1}\| &= \|\mathbf{\Lambda}_{T_1-T_2QT_3}\| = \|\mathbf{P}_{\mathcal{H}_{T_2}}(T_1 - T_2QT_3)|_{\mathcal{H}_{T_3}}\| \\ &\leq \|T_1 - T_2QT_3\|_\infty.\end{aligned}$$

Thus

$$\|\mathbf{\Lambda}_{T_1}\| \leq \inf_{Q\in\mathcal{H}^\infty_{n\times m}} \|T_1 - T_2QT_3\|_\infty.$$

For the other direction, we will need the commutant lifting theorem. We will prove the existence of Q_{opt}, such that

$$\|\mathbf{\Lambda}_{T_1}\| = \|T_1 - T_2Q_{opt}T_3\|_\infty.$$

This will complete the proof of the theorem. Accordingly let $\mathbf{S}_1$ and $\mathbf{S}_2$ denote multiplication by ζ on $\mathcal{L}^2_{n_3}$ and $\mathcal{L}^2_{n_2}$, respectively. Now $\mathcal{H}_{T_3}$ is invariant for $\mathbf{S}_1$ and $\mathcal{H}_{T_2}$ is invariant for $\mathbf{S}_2^*$. Set

$$\mathbf{U}_1 = \mathbf{S}_1|_{\mathcal{H}_{T_3}}, \quad \mathbf{U}_2 = \mathbf{P}_{\mathcal{H}_{T_2}}\mathbf{S}_2|_{\mathcal{H}_{T_2}}.$$

Then, $\mathbf{U}_1$ is an isometry, and $\mathbf{U}_2$ is a co-isometry. Moreover, it is easy to check that

$$\mathbf{P}_{\mathcal{H}_{T_2}}\mathbf{S}_2 = \mathbf{U}_2\mathbf{P}_{\mathcal{H}_{T_2}}, \quad \mathbf{S}_2\mathbf{M}_{T_1} = \mathbf{M}_{T_1}\mathbf{S}_1,$$

(where $\mathbf{M}_T$ denotes the multiplication operator by T and thus $\mathbf{\Lambda}_{T_1} = \mathbf{P}_{\mathcal{H}_{T_2}}\mathbf{M}_{T_1}|_{\mathcal{H}_{T_3}}$) and so

$$\mathbf{U}_2\mathbf{\Lambda}_{T_1} = \mathbf{\Lambda}_{T_1}\mathbf{U}_1.$$

Using the commutant lifting theorem, we have that there exists an operator $\mathbf{B} : \mathcal{L}^2_{n_3} \to \mathcal{L}^2_{n_2}$ such that

$$\begin{aligned}\|\mathbf{B}\| &= \|\mathbf{\Lambda}_{T_1}\|, \\ \mathbf{P}_{\mathcal{H}_{T_2}}\mathbf{B}|_{\mathcal{H}_{T_3}} &= \mathbf{\Lambda}_{T_1}, \\ \mathbf{S}_2\mathbf{B} &= \mathbf{B}\mathbf{S}_1.\end{aligned}$$

The last property implies that $\mathbf{B}$ is time-invariant, and so there exists $T \in \mathcal{L}^\infty_{n_3 \times n_2}$ with $\mathbf{B} = \mathbf{M}_T$. In particular,

$$\|T\|_\infty = \|\mathbf{B}\| = \|\mathbf{\Lambda}_{T_1}\|, \quad \mathbf{\Lambda}_{T_1} = \mathbf{\Lambda}_T.$$

But this means that $\mathbf{\Lambda}_{T_1 - T} = \mathbf{0}$, and so by a double application of the factorization argument used in the proof of Theorem 26, there exists $Q_{opt} \in \mathcal{H}^\infty_{n \times m}$ such that

$$T_1 - T = T_2 Q_{opt} T_3.$$

Hence

$$\|T_1 - T_2 Q_{opt} T_3\|_\infty = \|\mathbf{\Lambda}_{T_1}\|,$$

which completes the proof. □

8.2.4 Reduction to One Block Problem Setting

Computation of the largest singular values of the four block operator $\mathbf{A}$, defined in Section 8.2.1, and Young's operator $\mathbf{\Lambda}_{T_1}$, can be reduced to a one block problem by a series of spectral factorizations. In this section we discuss the problem of computing the largest singular value of $\mathbf{\Lambda}_{T_1}$.

First note that

$$\begin{aligned}
\mathcal{H}_{T_3} &= \mathcal{L}^2_{n_3} \ominus T_3^* \mathcal{K}^2_m = \mathcal{H}^2_{n_3} \oplus \mathcal{K}(T_3^*) \, , \\
\mathcal{H}_{T_2} &= \mathcal{L}^2_{n_2} \ominus T_2 \mathcal{H}^2_n = \mathcal{K}^2_{n_2} \oplus \mathcal{H}(T_2) \, ,
\end{aligned}$$

where

$$\begin{aligned}
\mathcal{K}(T_3^*) &:= \mathcal{K}^2_{n_3} \ominus T_3^* \mathcal{K}^2_m \, , \\
\mathcal{H}(T_2) &:= \mathcal{H}^2_{n_2} \ominus T_2 \mathcal{H}^2_n \, .
\end{aligned}$$

On the other hand, γ is a singular value of $\mathbf{\Lambda}_{T_1}$ if and only if there exists a non-zero $x \in \mathcal{H}_{T_3}$ satisfying

$$(\gamma^2 \mathbf{I} - \mathbf{\Lambda}^*_{T_1} \mathbf{\Lambda}_{T_1}) x = 0,$$

which is equivalent to

$$(\gamma^2\mathbf{I} - \mathbf{P}_{\mathcal{H}_{T_3}} T_1(\mathbf{U})^* \mathbf{P}_{\mathcal{H}_{T_2}} T_1(\mathbf{U}))x = 0 \ , \tag{8.16}$$

where $\mathbf{P}_{\mathcal{H}}$ denotes the orthogonal projection onto the subspace $\mathcal{H}$. Using the above decomposition of $\mathcal{H}_{T_3}$ we see that the singular vector candidate x should be in the form

$$x = x_+ + x_-, \qquad \text{where} \qquad x_+ \in \mathcal{H}^2_{n_3}, \qquad x_- \in \mathcal{K}(T_3^*).$$

Taking the projections of (8.16) onto $\mathcal{H}^2_{n_3}$ and onto $\mathcal{K}^2_{n_3}$ we get

$$\begin{aligned} 0 &= (\gamma^2\mathbf{I} - \mathbf{\Gamma}_2^*\mathbf{\Gamma}_2)x_+ - \mathbf{\Gamma}_2^*\mathbf{\Upsilon}_{23}x_- \ , \\ 0 &= (\gamma^2\mathbf{I} - \mathbf{\Gamma}_3\mathbf{\Gamma}_3^* - \mathbf{\Upsilon}_{23}^*\mathbf{\Upsilon}_{23})x_- - \mathbf{\Upsilon}_{23}^*\mathbf{\Gamma}_2 x_+, \end{aligned}$$

where we have used the notation

$$\begin{aligned} \mathbf{\Gamma}_2 &:= \mathbf{P}_{\mathcal{H}(T_2)} T_1(\mathbf{U})|_{\mathcal{H}^2_{n_3}} : \ \mathcal{H}^2_{n_3} \to \mathcal{H}(T_2) \ , \\ \mathbf{\Gamma}_3 &:= \mathbf{P}_{\mathcal{K}(T_3^*)} T_1(\mathbf{U})^*|_{\mathcal{K}^2_{n_2}} \ : \ \mathcal{K}^2_{n_2} \to \mathcal{K}(T_3^*) \ , \\ \mathbf{\Upsilon}_{23} &:= \mathbf{P}_{\mathcal{H}(T_2)} T_1(\mathbf{U})|_{\mathcal{K}(T_3^*)} \ : \ \mathcal{K}(T_3^*) \to \mathcal{H}(T_2). \end{aligned}$$

In summary γ is a singular value of $\mathbf{\Lambda}_{T_1}$ if and only if there exist $x_+ \in \mathcal{H}^2_{n_3}$ and $x_- \in \mathcal{K}(T_3^*)$ not both zero, such that the following holds

$$\left(\begin{bmatrix} \gamma^2\mathbf{I} & 0 \\ \mathbf{0} & \gamma^2\mathbf{I} - \mathbf{\Gamma}_3\mathbf{\Gamma}_3^* \end{bmatrix} - \begin{bmatrix} \mathbf{\Gamma}_2^* \\ \mathbf{\Upsilon}_{23}^* \end{bmatrix} [\mathbf{\Gamma}_2 \ \ \mathbf{\Upsilon}_{23}] \right) \begin{bmatrix} x_+ \\ x_- \end{bmatrix} = \begin{bmatrix} 0 \\ 0 \end{bmatrix} . \tag{8.17}$$

In the special case where T_3 is square and constant, i.e., it can be taken as identity, we have $\mathcal{K}(T_3^*) = \{0\}$, so $\mathbf{\Gamma}_3 = \mathbf{0}$ and $\mathbf{\Upsilon}_{23} = \mathbf{0}$. Hence, in this case (8.17) reduces to $x_- = 0$ and $x_+ \neq 0$ satisfies

$$(\gamma^2\mathbf{I} - \mathbf{\Gamma}_2^*\mathbf{\Gamma}_2)x_+ = 0. \tag{8.18}$$

Similarly, when T_2 is square and constant we have $\mathcal{H}(T_2) = \{0\}$, which implies that $\mathbf{\Gamma}_2 = \mathbf{0}$ and $\mathbf{\Upsilon}_{23} = \mathbf{0}$. Therefore, in this case (8.17) is equivalent to $x_+ = 0$ and

$$(\gamma^2\mathbf{I} - \mathbf{\Gamma}_3\mathbf{\Gamma}_3^*)x_- = 0. \tag{8.19}$$

Since $\mathbf{\Gamma}_3^* = \mathbf{P}_{\mathcal{K}^2_{n_2}} \mathbf{\Lambda}_{T_1}|_{\mathcal{K}(T_3^*)}$ we have $\|\mathbf{\Lambda}_{T_1}\| = \gamma_{opt} \geq \|\mathbf{\Gamma}_3\|$. If $\gamma_{opt} = \|\mathbf{\Gamma}_3\|$, then γ_{opt} is the solution of a one block problem. Therefore, we can assume $\gamma > \|\mathbf{\Gamma}_3\|$. In this case we can find a spectral factor $\mathbf{V}_\gamma$ which is invertible on $\mathcal{K}^2_{n_3}$ and satisfies

$$\mathbf{V}_\gamma \mathbf{V}_\gamma^* = \gamma^2 \mathbf{I} - \mathbf{\Gamma}_3 \mathbf{\Gamma}_3^*.$$

Then, defining $\hat{x}_- := \mathbf{V}_\gamma^* x_-$, the singular value/singular vector equation (8.17) can be re-written as

$$\left(\begin{bmatrix} \mathbf{I} & 0 \\ 0 & \mathbf{I} \end{bmatrix} - \begin{bmatrix} \gamma^{-1}\mathbf{\Gamma}_2^* \\ \mathbf{V}_\gamma^{-1}\mathbf{\Upsilon}_{23}^* \end{bmatrix} \begin{bmatrix} \gamma^{-1}\mathbf{\Gamma}_2 & \mathbf{\Upsilon}_{23}\mathbf{V}_\gamma^{*^{-1}} \end{bmatrix} \right) \begin{bmatrix} \gamma x_+ \\ \hat{x}_- \end{bmatrix} = \begin{bmatrix} 0 \\ 0 \end{bmatrix}. \quad (8.20)$$

Therefore, 1 must be a singular value of $[\gamma^{-1}\mathbf{\Gamma}_2 \quad \mathbf{\Upsilon}_{23}\mathbf{V}_\gamma^{*^{-1}}]$ or equivalently of its adjoint. In other words, $\gamma > \|\mathbf{\Gamma}_3\|$ is a singular value of $\mathbf{\Lambda}_{T_1}$ if and only if there exists a non-zero $y \in \mathcal{H}(T_2)$ such that

$$(\mathbf{I} - \frac{1}{\gamma^2}\mathbf{\Gamma}_2\mathbf{\Gamma}_2^* - \mathbf{\Upsilon}_{23}(\gamma^2\mathbf{I} - \mathbf{\Gamma}_3\mathbf{\Gamma}_3^*)^{-1}\mathbf{\Upsilon}_{23}^*)\, y = 0. \quad (8.21)$$

By studying the left hand side of (8.21) one can find a set of necessary and sufficient conditions for γ to be a singular value of $\mathbf{\Lambda}_{T_1}$. Note that the left hand side of the above equation is an operator acting on an element of $\mathcal{H}(T_2)$ and producing an element in the same subspace of $\mathcal{H}^2_{n_2}$. This operator can be expressed in terms of multiplication operators, by T_1 and T_1^*, and orthogonal projection operators $\mathbf{P}_{\mathcal{K}(T_3^*)}$ and $\mathbf{P}_{\mathcal{H}(T_2)}$. The action of each of these operators on an element in their respective domains can be explicitly computed, as in the SISO case. This is the typical functional setting for one block problems to which the skew Toeplitz methods apply. Although the operator in (8.21) is not a skew Toeplitz operator, the geometrical functional framework is the same.

Moreover, when T_2 and T_3 are transfer matrices with rational entries, the subspaces $\mathcal{H}(T_2)$ and $\mathcal{K}(T_3^*)$ are finite dimensional. Hence, the left hand side of (8.21) can be computed in terms of finite size matrices (representations of projection operators on the finite dimensional spaces). In the next section, we will study an infinite dimensional two block example to give an idea about the procedure for the general case.

8.3 MIMO Two Block Problem

In this section we present a solution to MIMO two block problem for stable distributed parameter plants satisfying certain commutativity assumptions. We note that due to the reduction method sketched above, the techniques of this section can be applied to more general problems.

A typical example of the two block problem is the mixed sensitivity minimization: given two weights W_1, W_2 and the plant P find

$$\gamma_{opt} = \inf_{C \text{ stabilizing } P} \left\| \begin{bmatrix} W_1(I+PC)^{-1} \\ W_2PC(I+PC)^{-1} \end{bmatrix} \right\|_\infty . \tag{8.22}$$

We will assume that the plant is stable, and it has the same number of inputs and outputs, i.e. $P \in \mathcal{H}^\infty_{n\times n}(\mathbb{C}_+)$. Then, all stabilizing controllers are in the form

$$C = Q(I-PQ)^{-1}, \qquad Q \in \mathcal{H}^\infty_{n\times n}(\mathbb{C}_+). \tag{8.23}$$

In the rest of this chapter we will drop the subscript $n \times n$, and the right half plane indicator $(\mathbb{C}_+)$ whenever the dimensions are clear from the context. Inserting the parametrization (8.23) in (8.22) we see that

$$\gamma_{opt} = \inf_{Q\in\mathcal{H}^\infty} \left\| \begin{bmatrix} W_1 \\ 0 \end{bmatrix} - \begin{bmatrix} W_1 \\ -W_2 \end{bmatrix} PQ \right\|_\infty . \tag{8.24}$$

Now, in order to avoid inner outer factorizations involving irrational transfer function matrices, we make some structural commutativity assumptions.

Assumption 8.1: The plant is in the form $P = M_1P_fP_o$ where M_1 is an $n \times n$ inner matrix, P_f is a rational $n \times n$ stable transfer function, and P_o is an $n \times n$ outer matrix invertible in $\mathcal{H}^\infty_{n\times n}$. Here M_1 and P_o are possibly infinite dimensional. Also assume that W_1, W_2, and M_1 are diagonal matrices, and that W_1 and W_2 are rational.

For example, if W_1 and W_2 are diagonal matrices, systems with output delays, whose general structure is shown in Figure 8.2, satisfy this

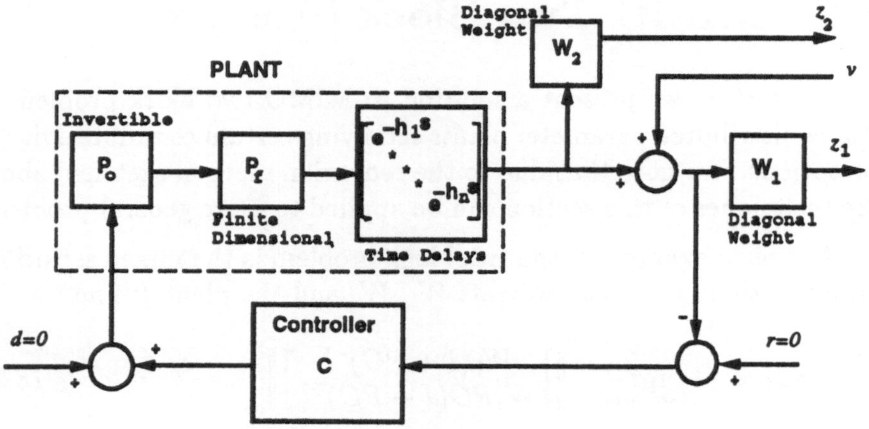

Figure 8.2: MIMO System With Output Delays

assumption (in this case M_1 is a diagonal matrix whose kth diagonal entry is $e^{-h_k s}$ for $h_k \geq 0$, $k = 1, \ldots, n$). In Figure 8.2, v is the output disturbance, and $z = [z_1^T, z_2^T]^T$ is the weighted output to be regulated. The problem of finding

$$\inf_{C \text{ stabilizing } P} \sup_{v \neq 0} \frac{\|z\|_2}{\|v\|_2}$$

is equivalent to finding γ_{opt} in (8.22). A numerical example would be:

$$M_1 = \begin{bmatrix} e^{-h_1 s} & 0 \\ 0 & e^{-h_2 s} \end{bmatrix}, \quad P_f = \begin{bmatrix} \frac{s-1}{s^2+0.5s+2} & \frac{1}{s+1} \\ 0 & \frac{2}{s+3} \end{bmatrix},$$

$$P_o = \begin{bmatrix} 1 - \frac{e^{-\tau_1 s}}{2s+5} & 0 \\ \frac{s-2}{s+7} & (1 + \frac{e^{-\tau_2 s}}{3s+2})^{-1} \end{bmatrix}, \quad W_1(s) = \begin{bmatrix} \frac{1}{2s+1} & 0 \\ 0 & \frac{s+1}{4s+1} \end{bmatrix},$$

and $W_2(s) = (s+2)I_{2\times 2}$, where h_1, h_2, τ_1, τ_2 are positive constants.

We now return to the problem defined in (8.24). The first step is to perform a spectral factorization on the weights: By Assumption 8.1, find a diagonal rational matrix G such that G^{-1} is outer, and satisfies

$$G^* G = W_1^* W_1 + W_2^* W_2.$$

Clearly, M_1 commutes with G. Then, (8.24) can be re-written as

$$\gamma_o = \inf_{Q \in \mathcal{H}^\infty} \left\| \begin{bmatrix} W_1 \\ 0 \end{bmatrix} - \begin{bmatrix} W_1 G^{-1} \\ -W_2 G^{-1} \end{bmatrix} M_1 G P_f P_o Q \right\|_\infty . \tag{8.25}$$

Assumption 8.2 The $n \times n$ matrix GP_f is bi-proper. That is, it has an inner-outer factorization of the form

$$GP_f = M_2 P_2$$

where $M_2, P_2, P_2^{-1} \in \mathcal{H}^\infty_{n\times n}$ and M_2 is inner.

This assumption is quite standard in $\mathcal{H}^\infty$ control. It is required for the properness of the optimal controller. The numerical example given above satisfies it. Clearly, it also implies that M_2 is rational. Recall that a similar assumption has been made for the SISO case. With the notation used above, we see that the (8.24) problem is equivalent to

$$\gamma_{opt} = \inf_{Q_1 \in \mathcal{H}^\infty} \|T_1 - T_2 Q_1\|_\infty$$

where $Q_1 = P_2 P_o Q$, and

$$T_1 = \begin{bmatrix} W_1 \\ 0 \end{bmatrix}, \quad T_2 = \begin{bmatrix} W_1 G^{-1} \\ -W_2 G^{-1} \end{bmatrix} M_1 M_2 .$$

Note that T_2 is inner and there is an invertible relationship between Q_1 and Q.

Now, from Theorem 27 we have $\gamma_{opt} = \|\mathbf{\Lambda}_{T_1}\|$; and in our case $T_3 = I$, $n_3 = n$, $m = n$, and $n_2 = 2n$. So, from the discussion of the previous section we see that $\mathbf{\Gamma}_3 = \mathbf{0}$ and $\mathbf{\Upsilon}_{23} = \mathbf{0}$. Therefore, γ is a singular value of $\mathbf{\Lambda}_{T_1}$ if and only if there exists a non-zero $x_+ \in \mathcal{H}^2_n$ such that

$$(\gamma^2 \mathbf{I} - \mathbf{\Gamma}_2^* \mathbf{\Gamma}_2) x_+ = 0 \tag{8.26}$$

where $\mathbf{\Gamma}_2 = \mathbf{P}_{\mathcal{H}(T_2)} T_1(\mathbf{S})|_{\mathcal{H}^2_n}$, which is an operator from $\mathcal{H}^2_n$ to $\mathcal{H}(T_2)$. By using T_1, T_2 given above, and the definition $\mathcal{H}(T_2) := \mathcal{H}^2_{2n} \ominus T_2\mathcal{H}^2_n$, we see that (8.26) can be written as follows

$$\mathbf{P}_{\mathcal{H}^2_n}\{\gamma^2 I - W_1^* W_1 + W_1^* W_1 G^{-1} M_1 M_2 \mathbf{P}_{\mathcal{H}^2_n} M_2^* M_1^* G^{*^{-1}} W_1^* W_1\} x_+ = 0 \,;$$

(here and in the sequel we simply write W_1, M_1, etc., for the operators $W_1(\mathbf{U})$, $M_1(\mathbf{U})$, etc.) which is equivalent to

$$\mathbf{P}_{\mathcal{H}^2_n}\Big(\gamma^2 I - W_1^*(I - W_1 G^{-1} G^{*^{-1}} W_1^*) W_1$$

$$-W_1^* W_1 G^{-1} M_1 M_2 \mathbf{P}_{\mathcal{K}^2_n} M_2^* M_1^* G^{*^{-1}} W_1^* W_1\Big)\, x_+ = 0.$$

It is interesting to note that $T_{2\perp} T_{2\perp}^* = I - T_2 T_2^*$ (see Section 8.1) and

$$W_1^*(I - W_1 G^{-1} G^{*^{-1}} W_1^*) W_1 = (T_1^* T_{2\perp})(T_{2\perp}^* T_1)$$

and from (8.2) it follows that $\gamma_{opt} \geq \|T_{2\perp}^* T_1\|_\infty$. If $\gamma_{opt} = \|T_{2\perp}^* T_1\|_\infty$, then the computation is completed. Therefore, γ, which is our candidate for the largest singular value of $\mathbf{\Lambda}_{T_1}$, can be taken as $\gamma > \|T_{2\perp}^* T_1\|_\infty$. Hence, for such a choice of γ there exists a diagonal $U_\gamma \in \mathcal{H}^\infty_{n\times n}$ with $U_\gamma^{-1} \in \mathcal{H}^\infty_{n\times n}$ such that

$$U_\gamma^* U_\gamma = \gamma^2 I - W_1^*(I - W_1 G^{-1} G^{*^{-1}} W_1^*) W_1.$$

This means that (8.26) holds for a non-zero $x_+ \in \mathcal{H}^2_n$ if and only if there exists a non-zero $y_+ \in \mathcal{H}^2_n$ satisfying

$$(\mathbf{I} - \mathbf{P}_{\mathcal{H}^2_n} U_\gamma^{*^{-1}} W_1^* W_1 G^{-1} M_1 M_2 \mathbf{P}_{\mathcal{K}^2_n} M_2^* M_1^* G^{*^{-1}} W_1^* W_1 U_\gamma^{-1}) y_+ = 0. \quad (8.27)$$

This y_+ is connected to x_+ by $U_\gamma x_+ = y_+$. Define

$$\Theta_\gamma := M_2^* M_1^* G^{*^{-1}} W_1^* W_1 U_\gamma^{-1}.$$

Then note that (8.27) is a singular value/singular vector equation for the Hankel operator Γ_{Θ_γ} whose symbol is Θ_γ, i.e. it is in the form

$$(\mathbf{I} - \Gamma^*_{\Theta_\gamma}\Gamma_{\Theta_\gamma})\, y_+ = 0. \tag{8.28}$$

Some important remarks are in order now. First note that $\Theta^*_\gamma\Theta_\gamma$ is a rational function determined by W_1, W_2 and γ. Let us denote a diagonal spectral factor of $\Theta^*_\gamma\Theta_\gamma$ by $W \in \mathcal{H}^\infty_{n\times n}$, i.e. W is a square outer rational matrix which satisfies

$$\Theta^*_\gamma\Theta_\gamma = U^{*^{-1}}_\gamma W^*_1 W_1 G^{-1} G^{*^{-1}} W^*_1 W_1 U^{-1}_\gamma = W^* W$$

(obviously W depends also on γ). The second important point to note is that, by Assumption 8.1, M_1 commutes with W_1, W^*_1, G^{-1}, $G^{*^{-1}}$ U^{-1}_γ and $U^{*^{-1}}_\gamma$, so we have

$$\Theta_\gamma\Theta^*_\gamma = M^*_2 G^{*^{-1}} W^*_1 W_1 U^{-1}_\gamma U^{*^{-1}}_\gamma W^*_1 W_1 G^{-1} M_2$$

which is rational. Let us denote the spectral factor of $\Theta_\gamma\Theta^*_\gamma$ by $W_0 \in \mathcal{H}^\infty_{n\times n}$, i.e. W_0 is a square rational outer matrix which satisfies

$$\Theta_\gamma\Theta^*_\gamma = W_0 W^*_0.$$

Again, W_0 depends on γ. It is not hard to see that these finite dimensional spectral factorizations for $\Theta_\gamma\Theta^*_\gamma$ and for $\Theta^*_\gamma\Theta_\gamma$ imply the following form for Θ_γ

$$\Theta_\gamma = W_0 M^* = M^*_0 W\ , \tag{8.29}$$

where $M = M_1\widehat{M}_2$ and $M_0 = \widetilde{M}_2 M_1 M_2$ for some rational square inner matrices $\widehat{M}_2$ and $\widetilde{M}_2$ obtained from W_1, and W_2 by a series of finite dimensional spectral factorizations. It is worth mentioning once more that in these representations of Θ_γ the square outer matrices W and W_0 are finite dimensional, and the square inner matrices M and M_0 are possibly infinite dimensional.

In the rest of this section, we will investigate necessary and sufficient conditions for a non-zero $y_+ \in \mathcal{H}^2_n$ to satisfy (8.28). We will use the factorization (8.29) for Θ_γ. Our purpose is to obtain a set of finitely many linear equations for γ to be a singular value of $\mathbf{\Lambda}_{T_1}$, (or equivalently for 1 to be a singular value of $\mathbf{\Gamma}_{\Theta_\gamma}$). The derivation of these finitely many linear equations is very similar to the computations of Chapter 4. In order to give the reader an idea on how these computations can be done directly in the $s-$domain, in this section we prefer not to transform the problem data to the $z-$domain.

Recall that we are dealing with the following singular value/singular vector equation

$$(\mathbf{P}_{\mathcal{H}^2_n} M W_0^* \mathbf{P}_{\mathcal{K}^2_n} W_0 M^* - \mathbf{I}) y_+ = 0. \tag{8.30}$$

As usual we first decompose y_+ into two orthogonal parts: $y_+ = u + Mv$, where $u \in \mathcal{H}(M)$ and $v \in \mathcal{H}^2_n$. Note that in the multivariable case $\mathcal{H}(M)$ is defined as $\mathcal{H}^2_n \ominus M\mathcal{H}^2_n$. With this decomposition, (8.30) becomes equivalent to

$$\begin{aligned} &\mathbf{P}_{\mathcal{H}^2_n} M W_0^* W_0 M^* u - M W_0^* \mathbf{P}_{\mathcal{H}^2_n} W_0 M^* u \\ &\qquad + \mathbf{P}_{\mathcal{K}^2_n} M W_0^* \mathbf{P}_{\mathcal{H}^2_n} W_0 M^* u = u + Mv. \end{aligned} \tag{8.31}$$

Recall that by (8.29) we have

$$M W_0^* W_0 M^* = W^* W. \tag{8.32}$$

Now applying $\mathbf{P}_{\mathcal{H}^2_n} M^*$ and $(\mathbf{I} - \mathbf{P}_{\mathcal{H}^2_n} M^*)$ to both sides of (8.31) we separate it into two equations defined on two orthogonal subspaces of $\mathcal{L}^2_n$. The first one is

$$\begin{aligned} v &= \mathbf{P}_{\mathcal{H}^2_n} W_0^* W_0 (M^* u) - \mathbf{P}_{\mathcal{H}^2_n} W_0^* \mathbf{P}_{\mathcal{H}^2_n} W_0 (M^* u) \\ &= \mathbf{P}_{\mathcal{H}^2} W_0^* \mathbf{P}_{\mathcal{K}^2} W_0 M^* u = 0 \, , \end{aligned} \tag{8.33}$$

and the second one is

$$(W_0^* W_0 - I)(M^* u) = \mathbf{P}_{\mathcal{H}^2_n} W_0^* W_0 (M^* u) + \mathbf{P}_{\mathcal{K}^2_n} W_0^* \mathbf{P}_{\mathcal{H}^2_n} W_0 (M^* u)$$

$$+M^*\mathbf{P}_{\mathcal{K}_n^2}W^*Wu - M^*\mathbf{P}_{\mathcal{K}_n^2}MW_0^*\mathbf{P}_{\mathcal{H}_n^2}W_0(M^*u). \qquad (8.34)$$

Note that if we multiply (8.34) by M on the left, we get

$$(W^*W-I)u = M\mathbf{P}_{\mathcal{H}_n^2}W_0^*W_0(M^*u) + M\mathbf{P}_{\mathcal{K}_n^2}W_0^*\mathbf{P}_{\mathcal{H}_n^2}W_0(M^*u)$$

$$+\mathbf{P}_{\mathcal{K}_n^2}W^*Wu - \mathbf{P}_{\mathcal{K}_n^2}MW_0^*\mathbf{P}_{\mathcal{H}_n^2}W_0(M^*u). \qquad (8.35)$$

We can compute the singular vector $y_+ = u$ from (8.34,8.35). Note that since $u \in \mathcal{H}(M)$ we have that $u_\perp := M^*u \in \mathcal{K}_n^2$. Therefore, all the projections on the right hand sides of (8.34,8.35) are finite rank. In other words, these projections can be explicitly computed in terms of $\phi_{-n_o}, \ldots, \phi_{-1}, \phi_1, \ldots, \phi_{n_1}$, where $\phi_{-k} = u_\perp(p_k)$, and $\phi_j = u(-r_j)$. Here p_k and r_j denote the poles of $W_0(s)$ and $W(s)$ for $k = 1, \ldots, n_o$ and $j = 1, \ldots, n_1$, respectively. We assume that p_k's and r_j's are distinct, and write

$$\begin{aligned} W_0(s) &= \frac{W_{p_1}}{s-p_1} + \cdots + \frac{W_{p_{n_o}}}{s-p_{n_o}}, \\ W(s) &= \frac{X_{r_1}}{s-r_1} + \cdots + \frac{X_{r_{n_1}}}{s-r_{n_1}} \end{aligned}$$

As in Chapter 4 let us define

$$\Phi_- = [\phi_{-1}^T \ldots, \phi_{-n_o}^T]^T \quad \text{and} \quad \Phi_+ = [\phi_1^T, \ldots, \phi_{n_1}^T]^T.$$

Then, the projections on the right hand sides of (8.34,8.35) can be computed explicitly in terms of Φ_- and Φ_+, and these equations can be written in the form:

$$u_\perp(s) = (W_0^T(-s)W_0(s) - I)^{-1}[R_1(s) \;\; R_2(s)] \begin{bmatrix} \Phi_- \\ \Phi_+ \end{bmatrix} \qquad (8.36)$$

and

$$u(s) = (W^T(-s)W(s) - I)^{-1}M(s)[R_1(s) \;\; R_2(s)] \begin{bmatrix} \Phi_- \\ \Phi_+ \end{bmatrix} \qquad (8.37)$$

where $R_1(s)$ and $R_2(s)$ are matrices that are computed from the finite rank projections:

$$R_1(s)\ \Phi_- = \mathbf{P}_{\mathcal{H}^2_n} W_0^* W_0 u_\perp + \mathbf{P}_{\mathcal{K}^2_n} W_0^* \mathbf{P}_{\mathcal{H}^2_n} W_0 u_\perp$$

$$-M^* \mathbf{P}_{\mathcal{K}^2_n} M W_0^* \mathbf{P}_{\mathcal{H}^2_n} W_0 u_\perp\ , \qquad (8.38)$$

$$R_2(s)\ \Phi_+ = M^* \mathbf{P}_{\mathcal{K}^2_n} W^* W u. \qquad (8.39)$$

The reader can check that the right hand sides of (8.38) and (8.39) are given by:

$$\begin{aligned} R_1(s)\ \Phi_- &= \sum_{k=1}^{n_o} W_0^T(-p_k)\frac{W_{p_k}}{s-p_k}\phi_{-k} + \sum_{k=1}^{n_o}\frac{-W_{p_k}^T}{s+p_k}\sum_{j=1}^{n_o}\frac{-W_{p_j}}{p_k+p_j}\phi_{-j} \\ &\quad - M^T(-s)\sum_{k=1}^{n_o} M(-p_k)\frac{-W_{p_k}^T}{s+p_k}\sum_{j=1}^{n_o}\frac{-W_{p_j}}{p_k+p_j}\phi_{-j}\ , \qquad (8.40) \\ R_2(s)\ \Phi_+ &= M^T(-s)\sum_{j=1}^{n_1}\frac{-X_{r_j}^T}{s+r_j}W(-r_j)\phi_j\ . \qquad (8.41) \end{aligned}$$

Note that $(W_0^* W_0 - I)^{-1}$ and $(W^* W - I)^{-1}$ can be expanded as follows

$$\begin{aligned} (W_0^T(-s)W_0(s) - I)^{-1} &=: D_0 + \sum_{k=1}^{n_o}\frac{Y_{\alpha_k}}{s-\alpha_k} - \frac{Y_{\alpha_k}^T}{s+\alpha_k} \\ (W^T(-s)W(s) - I)^{-1} &=: D + \sum_{j=1}^{n_1}\frac{Z_{\beta_j}}{s-\beta_j} - \frac{Z_{\beta_j}^T}{s+\beta_j}, \end{aligned}$$

where $\alpha_1,\ldots,\alpha_{n_o},\beta_1,\ldots,\beta_{n_1} \in \overline{\mathbb{C}_+}$. Now it is easy to see that (8.36) leads to the following necessary conditions for $u_\perp$ to be in $\mathcal{K}^2_n$:

$$Y_{\alpha_k}\ (\ R_1(-\alpha_k)\Phi_- + R_2(-\alpha_k)\Phi_+\ ) = 0 \quad \forall\ k = 1,\ldots,n_o. \qquad (8.42)$$

Similarly from (8.37) we get the following necessary conditions for u to be in $\mathcal{H}^2_n$:

$$Z_{\beta_j}\ M(\beta_j)\ (\ R_1(\beta_j)\Phi_- + R_2(\beta_j)\Phi_+\ ) = 0 \quad \forall\ j = 1,\ldots,n_1. (8.43)$$

Assumption 8.3 The matrices Y_{α_k}, Z_{β_j} and $M(\beta_j)$ are invertible for all $k = 1, \ldots, n_o$ and $j = 1, \ldots, n_1$.

With this assumption, conditions (8.42) and (8.43) can be written as

$$\begin{bmatrix} \mathcal{E}_{11} & \mathcal{E}_{12} \\ \mathcal{E}_{21} & \mathcal{E}_{22} \end{bmatrix} \begin{bmatrix} \Phi_- \\ \Phi_+ \end{bmatrix} = \begin{bmatrix} 0 \\ 0 \end{bmatrix}, \tag{8.44}$$

where

$$\mathcal{E}_{11} = \begin{bmatrix} R_1(-\alpha_1) \\ \vdots \\ R_1(-\alpha_{no}) \end{bmatrix}, \quad \mathcal{E}_{12} = \begin{bmatrix} R_2(-\alpha_1) \\ \vdots \\ R_2(-\alpha_{no}) \end{bmatrix},$$

$$\mathcal{E}_{21} = \begin{bmatrix} R_1(\beta_1) \\ \vdots \\ R_1(\beta_{n1}) \end{bmatrix}, \quad \mathcal{E}_{22} = \begin{bmatrix} R_2(\beta_1) \\ \vdots \\ R_2(\beta_{n1}) \end{bmatrix}.$$

Assumption 8.4 The matrices $W_0(-p_k)$, W_{p_k}, $W(-r_j)$, X_{r_j} are invertible and α_k's and β_j's are distinct.

It is important to note that, with Assumption 8.4, the matrix $\mathcal{E}_{22}$ is invertible. Hence (8.44) can be reduced to a condition on Φ_- only:

$$(\mathcal{E}_{11} - \mathcal{E}_{12}\mathcal{E}_{22}^{-1}\mathcal{E}_{21})\Phi_- = 0. \tag{8.45}$$

Once Φ_- is computed, Φ_+ can be obtained from

$$\Phi_+ = -\mathcal{E}_{22}^{-1}\mathcal{E}_{21}\Phi_-.$$

Now we are ready to state the main result of this section.

Theorem 28 *Suppose that Assumptions 8.1–8.4 hold. Then, $\gamma > \|T_{21}^*{}_\perp T_1\|_\infty$ is a singular value of Λ_{T_1} if only if there exists a non-zero Φ_- such that*

$$\mathcal{R}_\gamma \Phi_- = 0 \quad \text{where} \quad \mathcal{R}_\gamma = \mathcal{E}_{11} - \mathcal{E}_{12}\mathcal{E}_{22}^{-1}\mathcal{E}_{21}. \tag{8.46}$$

Proof. The necessity of these conditions are immediate from the above discussion. As for the sufficiency, suppose we can find a non-zero Φ'_- satisfying $\mathcal{R}_\gamma \Phi'_- = 0$. Then we can define $\Phi'_+ = -\mathcal{E}_{22}^{-1}\mathcal{E}_{21}\Phi'_-$, and obtain u, resp. $u_\perp$, from these vectors via (8.37), resp. (8.36). Note that (8.42,8.43) and (8.36,8.37) imply that by construction $u \in \mathcal{H}_n^2$ and $u_\perp \in \mathcal{K}_n^2$. Now in order to complete the sufficiency part of the proof, we need to show that Φ_- (constructed from $u_\perp(p_k)$'s) and Φ_+ (constructed from $u(-r_j)$'s) correspond to Φ'_- and Φ'_+ respectively. Let us first consider (8.36) from which $u_\perp$ is constructed:

$$(W_0^* W_0 - I)u_\perp = R_1\Phi'_- + R_2\Phi'_+.$$

Applying $\mathbf{P}_{\mathcal{H}_n^2}$ on both sides of this equation we get

$$\sum_{k=1}^{n_o} W_0^T(-p_k)\frac{W_{p_k}}{s-p_k}\phi_{-k} = \sum_{k=1}^{n_o} W_0^T(-p_k)\frac{W_{p_k}}{s-p_k}\phi'_{-k}\,.$$

By the assumption that $W_0^T(-p_k)W_{p_k}$ is non-singular for all $k = 1, \ldots, n_o$, we have $\Phi_- = \Phi'_-$. Similarly, consider (8.37) from which u is defined:

$$(W^*W - I)u = M(R_1\Phi'_- + R_2\Phi'_+).$$

It is easy to see from (8.40,8.41) that by applying $\mathbf{P}_{\mathcal{K}_n^2}$ on both sides of this equation, we get

$$\sum_{j=1}^{n_1} \frac{-X_{r_j}^T}{s+r_j}W(-r_j)\phi_j = \sum_{j=1}^{n_1} \frac{-X_{r_j}^T}{s+r_j}W(-r_j)\phi_j{}'.$$

Again, by the assumption that $X_{r_j}^T W(-r_j)$ is non-singular for all $j = 1, \ldots, n_1$, we have $\Phi_+ = \Phi'_+$. This concludes the proof. □

Note that in the SISO case Theorem 28 gives the same equation which appears in Theorem 19 of Chapter 4. In the MIMO case, $\mathcal{R}_\gamma$ is of dimension $n_o n \times n_o n$. A Matlab code for the implementation of the above formula is presently being written.

Chapter 9

Notes and References

There are a huge number of references in $\mathcal{H}^\infty$ theory now in the literature. Thus this chapter will certainly not contain a complete survey, and will only concentrate on those specific references that were used in writing this book.

In this book we have presented an operator theoretic approach to several interesting control problems. We have also illustrated connections between these operator theoretic methods and Nevanlinna-Pick type of interpolation methods used in $\mathcal{H}^\infty$ control, and stability margin optimization problems.

A mathematical background is given in Chapter 2. Although we tried to make this chapter as self contained as possible, important details are left out. The reader may want to consult [28], [41], [56], [89], [94], [117], etc., for complete details.

In Chapters 2, 3, and 4, the interpolation theory results follow [96], [16], and [97]. A number of sections in these chapters are based on a tutorial review [78]. The reader should consult these papers for very rather complete lists of references on the subjects. The notion of "skew Toeplitz operator" was first defined in complete generality in [6]. See [11] for a computational solution to the gain-phase margin problem.

The main results of Chapter 5 are from [79]. Here we tried to make the presentation more reader friendly. In fact, here we used a spectral factorization to reduce the problem to a one block problem, so that the number of linear equations is reduced to $2(n+\ell)$, from $3n+2\ell$. This way the notation is simplified considerably.

The first part of Chapter 6 is from [77]. A brief summary of this section is also given in [78]. The second part of Chapter 6 is taken from [101]. This section presents alternative formulae for the optimal controller, as well as all suboptimal controllers. The controller expression presented in Chapter 6 section 2 is so simple that we were able to describe it in couple of pages using a notation independent of the earlier parts of the book. The proofs of these results are omitted here, see [101] and [52] for details. The proofs are based on AAK theory [1], and on the earlier results of [79] which are summarized in Chapter 5.

In Chapter 7 we present a flexible beam example from [67] and [68]. Here, we have chosen first order weights in order to simplify the computation of the $\mathcal{H}^\infty$ optimal performance and the controller. In fact, as the reader can check, the computations were carried out by hand. For a more realistic beam example, it may be necessary to consider higher order (possibly infinite dimensional) weights. Moreover, in the beam example considered here, we assume Kelvin-Voigt damping. Usually this under-estimates the damping for the modes at low frequency, then over-estimates it for higher frequency modes. There are different types of damping models which one can consider. But in essence the computations for $\mathcal{H}^\infty$ control would be similar.

The second part of Chapter 7 illustrates the computation of optimal controller for a delay system in the context of robustness optimization in the gap metric. For this example, we have used the controller expression given in Chapter 6. We should note that the same control problem has been studied in [43] and [44], where an equation similar to (7.30) has been obtained for computing the optimal robustness level and the optimal controller. Note that the controller expression (7.29) given here is simpler than the ones appearing in [43] and [44]. The same controller could be obtained by using the recent results of [19], where a state space approach has been taken. See also [72] and [95] for a time domain approach.

Chapter 8 is based on the work of [6], [22], [28], [33], [34], [39],

[58] and [81]. It is important to mention that for MIMO distributed plants, there are several $\mathcal{H}^\infty$ control problems which are still under investigation. Among recent results in this area are state space solutions (which involves operator valued Riccati equations) reported in [107], [106], [15]. See also [13], [14] and references therein. Also, in [102] the operator theoretic approach presented in this book has been combined with the one step extension theory of AAK [2], to give a solution to the problem of robustness optimization in the gap metric for a class of delay systems. The section on the commutant lifting theorem is based on [94] (Chapters I-II), and [28] (Chapter XIV). The four block operator was first defined in [22]. See also [34] and [81] for details about the spectral properties of this operator. Young's operator was first defined in [118]. See also [29] for a study of the uses of this operator in H^2–H^∞ suboptimization theory.

Bibliography

[1] Adamjan, V. M., D. Z. Arov, and M. G. Krein, "Analytic Properties of Schmidt Pairs for a Hankel Operator and the Generalized Shur–Takagi Problem," *Math. USSR Sbornik* **15** (1971), pp. 31–73.

[2] Adamjan, V. M., D. Z. Arov, and M. G. Krein, "Infinite Hankel block matrices and related problems," *AMS Translations* **111** (1978), pp. 133-156.

[3] Ball, J. A., and N. Cohen, "Sensitivity minimization in an H^∞ norm," *Int. J. Control* **46** (1986), pp. 785-816.

[4] Bart, H., I. Gohberg, M. A. Kaashoek, *Minimal Factorization of Matrix and Operator Functions*, Birkhäuser, Basel, 1979.

[5] Başar, T., and P. Bernhard, *H^∞ Optimal Control and Related Minimax Design Problems: a Dynamic Games Approach*, Birkhäuser, Boston, 1991.

[6] Bercovici, H., C. Foias, and A. Tannenbaum, "On skew Toeplitz operators," *Operator Theory: Advances and Applications* **32** (1988), pp. 21–43.

[7] Bontsema, J., R. F. Curtain, J. M. Schumacher, "Robust Control of Flexible Structures: a case study," *Automatica*, **24** (1988), pp. 177–186.

[8] Bontsema, J., and S. A. de Vries "Robustness of flexible structures against small time delays", *Proc. 27th IEEE Conf. Decision and Contr.*, 1988 pp. 1647-1648.

[9] Callier, F. M., and C. A. Desoer, "An algebra of transfer functions for distributed linear time-invariant systems," *IEEE Transactions on Circuits and Systems*, **25** (1978), pp. 651–662.

[10] Chen, M. J., and C. A. Desoer, "Necessary and sufficient condition for robust stability of linear distributed feedback systems," *Int. J. Control*, **35** (1982), pp. 255–267.

[11] Cockburn, J., Y. Sidar, and A. Tannenbaum, "A constructive solution to the gain-phase margin problem" *Proceedings of 31st IEEE Conference on Decision and Control*, Tucson AZ, December 1992, pp. 682–683.

[12] Curtain, R. F., and K. Glover, "Robust stabilization of infinite dimensional systems by finite dimensional controllers," *Systems and Control Letters*, **7** (1986), pp. 41–47.

[13] Curtain, R. F., "H^∞ control for distributed parameter systems: a survey," Proc. of the 29th CDC, Honolulu, Hawaii, December 1990, pp. 22–26.

[14] Curtain, R. F., "A synthesis of time and frequency domain methods for the control of infinite dimensional systems: a system theoretic approach," in *Control and Estimation in Distributed Parameter Systems*, H. T. Banks ed., SIAM Frontiers in Applied Mathematics vol. 11, 1992, pp. 171–224.

[15] Curtain, R. F., and A. J. Pritchard, "Robust stabilization of infinite dimensional systems with respect to coprime factor perturbations," technical report no. W-9220, Department of Mathematics, University of Groningen.

[16] Doyle, J., B. Francis, and A. Tannenbaum, *Feedback Control Theory*, Macmillan Publishing Co., New York, 1992.

[17] Doyle, J., K. Glover, P. P. Khargonekar, and B. Francis, "State space solutions to standard H^2 and H^∞ control problems," *IEEE Trans. Automatic Control*, **AC-34** (1989), pp. 831–847.

[18] Doyle, J. C., and G. Stein, "Multivariable feedback design: concepts for a classical/modern synthesis," *IEEE Transactions on Automatic Control*, **26** (1981), pp. 4–16.

[19] Dym, H., T. T. Georgiou, and M. C. Smith, "Direct design of optimal controllers for delay systems," *Proceedings of 32nd IEEE Conference on Decision and Control*, San Antonio TX, December 1993, pp. 3821–3823.

[20] Enns, D., H. Özbay, and A. Tannenbaum, "Abstract model and controller design for an unstable aircraft," *Journal of Guidance Control and Dynamics*, **15** (1992), pp. 498–508.

[21] Enns, D., H. Özbay, and A. Tannenbaum, "$\mathcal{H}^\infty$ optimal controllers for a distributed model of an unstable aircraft," Proc. of the 30th IEEE Conference on Decision and Control, Brighton, England, December 1991, pp. 3020–3024.

[22] Feintuch, A., and B. Francis, "Uniformly optimal control of linear systems," *Automatica* **21** (1986), pp. 563–574.

[23] Feintuch, A., and A. Tannenbaum, "Gain optimization for distributed plants," *Systems & Control Letters*, **6** (1986), pp. 295–301.

[24] Flamm, D. S., "Outer factor 'Absorption' for $\mathcal{H}^\infty$ control problems," *Int. J. Robust and Nonlinear Control*, **2** (1992), pp. 31–48.

[25] Flamm, D., and S. Mitter, "H^∞ Sensitivity Minimization for Delay Systems," *Systems and Control Letters*, **9** (1987), pp. 17–24.

[26] Flamm, D. S., H. Yang, Q. Ren, and K. Klipec, "Numerical computation of inner factors for distributed parameter

systems," ISS Report No. 58, Princeton University, Dept. of Electrical Engineering, August 1990; see also Proc. of the 29th Annual Allerton Conference on Communication, Control and Computation, Monticello IL, 1991, pp. 72–73.

[27] Flamm, D. S., and H. Yang, "Optimal mixed sensitivity for SISO distributed plants," *IEEE Transactions on Automatic Control*, **39** (1994), pp. 1150–1165.

[28] Foias, C., and A. E. Frazho, *The Commutant Lifting Approach to Interpolation Problems*, Birkhäuser, Basel, 1990.

[29] Foias, C., and A. Frazho, "Commutant lifting and simultaneous H^∞ and L^2 suboptimization," *SIAM J. Math. Anal.* **23** (1992), pp. 984–994

[30] Foias, C., and A. Tannenbaum, "On the Nehari problem for a certain class of L^∞ functions appearing in control theory," *J. of Functional Analysis* **74** (1987), pp. 146-159.

[31] Foias, C., and A. Tannenbaum, "On the Nehari problem for a certain class of L^∞ functions appearing in control theory, II" *J. of Functional Analysis* **81** (1988), pp. 207-218.

[32] Foias, C., and A. Tannenbaum, "Some remarks on optimal interpolation," *Systems and Control Letters*, **11** (1988), pp. 259–264.

[33] Foias, C., and A. Tannenbaum, "On the Four Block Problem, I," *Operator Theory: Advances and Applications* **32** (1988), pp. 93-112.

[34] Foias, C., and A. Tannenbaum, "On the four block problem, II: the singular system" *Operator Theory and Integral Equations* **11** (1988), pp. 726-767.

[35] Foias, C., and A. Tannenbaum, "On the parametrization of the suboptimal solutions in generalized interpolation," *Linear Algebra and its Applications*, **122/123/124** (1989), pp. 145–164.

[36] Foias, C., A. Tannenbaum, and G. Zames, "Weighted sensitivity minimization for delay systems," *IEEE Trans. Automatic Control*, **31** (1986), pp. 763–766.

[37] Foias, C., A. Tannenbaum, and G. Zames, "On the H^∞ optimal sensitivity problem for systems with delays," *SIAM J. Control and Optimization* **25** (1987), pp. 686-706.

[38] Foias, C., A. Tannenbaum, and G. Zames, "Some explicit formulae for the singular values of a certain Hankel operators with factorizable symbol," *SIAM J. Math. Analysis* **19** (1988), pp. 1081-1091.

[39] Francis, B., *A Course in H^∞ Control Theory*, Lecture Notes in Control and Information Sciences, vol. 88, Springer Verlag, 1987.

[40] Francis, B., G. Zames, "On $\mathcal{H}^\infty$ optimal sensitivity theory for SISO feedback systems," *IEEE Trans. Aut. Contr.*, **29** 1984, pp.9-16.

[41] Garnett, J. B., *Bounded Analytic Functions*, Academic Press, New York, 1981.

[42] Georgiou, T. T., and M. C. Smith, "Optimal robustness in the gap metric," *IEEE Trans. Automat. Control*, **35** (1990), pp. 673–686.

[43] Georgiou, T. T., and M. C. Smith, "Robust stabilization in the gap metric: controller design for distributed plants," *IEEE Transactions on Automatic Control*, **37** (1992), pp. 1133–1143.

[44] Georgiou, T. T., and M. C. Smith, "Topological approaches to robustness," in *Analysis and Optimization of Systems: State and Frequency Domain Approaches for Infinite Dimensional Systems*, R. F. Curtain Ed., Lecture Notes in Control and Information Sciences, vol. 185, Springer Verlag, 1993, pp. 222–241.

[45] Glover, K., "All optimal Hankel norm approximations of linear multivariable systems and their L^∞ error bounds," *Int. J. Control* **39** (1984), pp. 1115–1193.

[46] Glover, K., R. F. Curtain, and J. R. Partington, "Realisation and approximation of linear infinite dimensional systems with error bounds," *SIAM J. Control and Optimization*, **26** (1988), pp. 863–898.

[47] Glover, K., and J. C. Doyle, "State space formulae for all stabilizing controllers that satisfy an $\mathcal{H}^\infty$ norm bound and relations to risk sensitivity," *System and Control Letters*, **11** (1988), pp. 167–172.

[48] Glover, K., J. Lam, and J. R. Partington, "Balanced realisation and Hankel norm approximation of systems involving delays," Proc. of the 25th IEEE Conf. on Decision and Control, Athens, Greece, 1986, pp. 1810–1815.

[49] Glover, K., J. Lam, and J. R. Partington, "Rational approximation of a class of infinite dimensional systems, I: singular values of Hankel operators," *Mathematics of Control, Signals, and Systems*, **3** (1990), pp. 325–344.

[50] Glover, K., and D. McFarlane, "Robust stabilization of normalized coprime factor descriptions with $\mathcal{H}^\infty$ bounded uncertainty," *IEEE Trans. Automatic Control*, **34** (1989), pp. 821–830.

[51] Gu, C., "Eliminating the genericity conditions in the skew Toeplitz operator algorithm for $\mathcal{H}^\infty$ optimization," *SIAM J. Math. Anal.*, **23** (1992), pp. 1623–1636.

[52] Gu, C., O. Toker, and H. Özbay, "On the two-block $\mathcal{H}^\infty$ problem for a class of unstable distributed systems," *Linear Algebra and its Applications*, 1995, to appear.

[53] Gu, G., P. P. Khargonekar, and E. B. Lee, "Approximation of infinite dimensional systems," *IEEE Transactions on Automatic Control*, **34** (1989), pp. 610–618.

[54] Halmos, P. R., *A Hilbert Space Problem Book*, Springer Verlag, New York, 1982.

[55] Helton, J. W., "Broadbanding: gain equalization directly from data," *IEEE Trans. Circuits and Systems*, **28** (1981), pp. 1125–1137.

[56] Hoffman, K., *Banach Spaces of Analytic Functions*, Dover Publications, New York, reprint, originally published by Prentice Hall, 1962.

[57] Jonckheere, E., and M. Verma, "A spectral characterization of $\mathcal{H}^\infty$ optimal feedback performance and its efficient computation," *Systems and Control Letters* **8** (1985), pp. 13–22.

[58] Khargonekar, P. P., H. Özbay, and A. Tannenbaum, "The four block problem: stable plants and rational weights," *International Journal of Control*, **50** (1989), pp. 1013–1023.

[59] Khargonekar, P. P., and K. Poolla, "Robust stabilization of distributed systems," *Automatica*, **22** (1986), pp. 77–84.

[60] Khargonekar, P. P., and A. Tannenbaum, "Noneuclidean metrics and the robust stabilization of systems with parameter uncertainty," *IEEE Trans. Automatic Control*, **30** (1985), pp. 1005–1013.

[61] Kimura, H., "Robust stabilization for a class of transfer functions," *IEEE Trans. Automatic Control*, **29** (1984), pp. 788–793.

[62] Krein, M., and A. Nudelman, *The Markov Moment Problem and Extremal Problems*, AMS Publications, Providence RI, 1977.

[63] Kwakernaak, H., "Minimax frequency domain performance and robustness optimization of linear systems," *IEEE Trans. Automat. Control*, **30** (1985), pp. 994–1004.

[64] Le, D. K., and A. E. Frazho, "A numerical procedure for a non-rational $\mathcal{H}^\infty$ optimization problem in control design," *Systems & Control Letters*, **16** (1991), pp. 9–15.

[65] Lypchuk, T., M. C. Smith, and A. Tannenbaum, "Weighted sensitivity minimization: General plants in H^∞ and rational weights," *Linear Algebra Appl.*, **109** (1988), pp. 71–90.

[66] Lenz, K., H. Özbay, A. Tannenbaum, J. Turi, and B. Morton, "Robust control design for a flexible beam using a distributed parameter $\mathcal{H}^\infty$ method," Proc. of the 28th IEEE Conference on Decision and Control, Tampa FL, December 1989, pp. 2673–2678.

[67] Lenz, K., H. Özbay, A. Tannenbaum, J. Turi, and B. Morton, "Frequency domain analysis and robust control design for an ideal flexible beam," *Automatica*, **27** (1991), pp. 947–961.

[68] Lenz, K., and H. Özbay, "Analysis and robust control techniques for an ideal flexible beam," in *Multidisciplinary Engineering Systems: Design and Optimization Techniques and their Applications*, C. T. Leondes ed., Academic Press Inc., 1993, pp. 369–421.

[69] Mäkilä, P., "Approximation of stable systems by Laguerre filters," *Automatica*, **26** (1990), pp. 333–345.

[70] Morari, M., and E. Zafiriou, *Robust Process Control*, Prentice Hall, 1989.

[71] Morris, K., "Robustness of controllers designed using Galerkin type approximations," Proc. of the 28th CDC, Tampa Fl, 1989, pp. 2679–2684.

[72] Nagpal, K., and R. Ravi, "$\mathcal{H}^\infty$ control and estimation problems with delayed measurements: state space solutions," *Proc. of the American Control Conference*, Baltimore MD, June 1994, pp. 2379–2383.

[73] Nehari, Z., "On bounded bilinear forms," *Ann. of Math.*, **65** (1957), pp. 153–162.

[74] Nikolskii, N. K., *Treatise on the Shift Operator*, Springer Verlag, Berlin, 1986.

[75] Özbay, H., "A simpler formula for the singular values of a certain Hankel operator," *Systems and Control Letters*, **15** (1990), pp. 381–390.

[76] Özbay, H., "On the 1-block H^∞ control problem for a class of MIMO distributed systems," Proc. of 1991 American Control Conference, Boston MA, June 1991, pp. 1642–1647.

[77] Özbay, H., "Controller reduction in the 2-block H^∞ optimal design for distributed plants," *International Journal of Control*, **54** (1992), pp. 1291–1308.

[78] Özbay, H., "H^∞ optimal controller design for a class of distributed parameter systems," *Int. J. Control*, **58** (1993), pp. 739–782.

[79] Özbay, H., M. C. Smith, and A. Tannenbaum, "Mixed sensitivity optimization for a class of unstable infinite dimensional systems," *Linear Algebra and its Applications*, **178** (1993), pp. 43–83.

[80] Özbay, H., M. C. Smith, and A. Tannenbaum, "On the optimal two block $\mathcal{H}^\infty$ compensators for distributed unstable plants," Proc. of the 1992 American Control Conference, Chicago IL, June 1992, pp. 1865–1869.

[81] Özbay, H., and A. Tannenbaum, "A skew Toeplitz approach to the H^∞ optimal control of multivariable distributed systems," *SIAM J. Control and Optimization*, **28** (1990), pp. 653–670.

[82] Özbay, H., and A. Tannenbaum, "On the structure of suboptimal H^∞ controllers in the sensitivity minimization

problem for distributed stable systems," *Automatica*, **27** (1991), pp. 293–305.

[83] Özbay, H., and J. Turi, "On input/output stabilization of singular integrodifferential systems," *Applied Mathematics & Optimization*, **30** (1994), pp. 21–49.

[84] Partington, J. R., and K. Glover, "Robust stabilization of delay systems by approximation of coprime factors," *Systems & Control Letters*, **14** (1990), pp. 325–331.

[85] Partington, J. R., K. Glover, H. J. Zwart, and R. F. Curtain, "L_∞ approximation and nuclearity of delay systems," *Systems and Control Letters*, **10** (1988), pp. 59–65.

[86] Power, S. C., *Hankel Operators on Hilbert Space*, Pitman Advanced Publishing Program, Boston, 1982.

[87] Rodriguez, A., and M. A. Dahleh, "Weighted $\mathcal{H}^\infty$ optimization for stable infinite dimensional systems using finite dimensional techniques," Proc. of the 29th IEEE Conf. on Decision and Control, Honolulu HI, December 1990, pp. 1814–1819.

[88] Russell, D. L. "On mathematical models for the elastic beam with frequency-proportional damping," in *Control and Estimation in Distributed Parameter Systems*, H. T. Banks ed., SIAM Frontiers in Applied Mathematics vol. 11, 1992, pp. 125–169.

[89] Sarason, D., "Generalized interpolation in H^∞," *Transactions of the AMS*, **127** (1967), pp. 179–203.

[90] Smith, M. C., "Singular values and vectors of a class of Hankel operators," *System and Control Letters*, **12** (1989), pp. 301–308.

[91] Smith, M. C., "On stabilization and existence of coprime factorizations," *IEEE Transactions on Automatic Control*, 1989, pp. 1005–1007.

[92] Smith, M. C., "Well-posedness of $\mathcal{H}^\infty$ optimal control problems," *SIAM J. Control and Optimization*, **28** (1990), pp. 342–358.

[93] Sz.-Nagy, B., and C. Foias, "Dilation des commutants d'opérateurs," *C. R. Acad. Sci. Paris*, Serie A, **266** (1968), pp. 493–495.

[94] Sz.-Nagy, B., and C. Foias, *Harmonic Analysis of Operators on Hilbert Space*, North Holland, Amsterdam, 1970.

[95] Tadmor, G., "H^∞ control in systems with a single input delay," *Proc. of the American Control Conference*, Seattle WA, June 1995, pp. 321–325.

[96] Tannenbaum, A., *Invariance and System Theory: Algebraic and Geometric Aspects*, **845**, Springer-Verlag, New York, 1981.

[97] Tannenbaum, A., "Frequency domain methods for the $\mathcal{H}^\infty$-optimization of distributed systems," in *Analysis and Optimization of Systems: State and Frequency DOmain Approaches for Infinite Dimensional Systems*, Curtain, Bensoussan and Lions eds., Springer-Verlag, Berlin 1992; Proc. of the 10th International Conference, Sophia-Antipolis, France, June 1992, pp. 242–275.

[98] Toker, O., and H. Özbay, "On the computation of suboptimal $\mathcal{H}^\infty$ controllers for unstable infinite dimensional systems," in *Robust Control Theory*, B. A. Francis and P. P. Khargonekar Eds., IMA Volumes in Mathematics and its Applications, Vol. 66, pp. 105–128, Springer-Verlag, New York, 1995.

[99] Toker, O., and H. Özbay, "On suboptimal $\mathcal{H}^\infty$ controllers for unstable distributed plants," *Proc. of the American Control Conference*, San Francisco CA, June 1993, pp. 2160–2164.

[100] Toker, O., and H. Özbay, "Numerical issues in the computation of H^∞ controllers for distributed plants," *Proceedings of 31st Allerton Conference on Communication, Control and Computing*, University of Illinois Urbana-Champaign, September-October 1993, pp. 535–544.

[101] Toker, O., and H. Özbay, "$\mathcal{H}^\infty$ optimal and suboptimal controllers for infinite dimensional SISO plants," *IEEE Transactions on Automatic Control*, **40** (1995) pp. 751–755.

[102] Toker, O., and H. Özbay, "Gap metric problem for MIMO delay systems: parameterization of all suboptimal controllers," to appear in *Automatica*; a brief version appears under the title "Suboptimal robustness in the gap metric for MIMO delay systems," in *Proceedings of the American Control Conference*, Baltimore MD, June-July 1994, pp. 3183–3187.

[103] Tu, H., *An $\mathcal{H}^\infty$ Optimization Method and Matlab Program for Linear Distributed Systems*, M.S. Thesis, Department of Mathematics and Statistics, University of Minnesota at Duluth, September 1992.

[104] Ulus, C., *Control and Approximation of Delay Systems*, M.S. Thesis, Department of Electrical Engineering, The Ohio State University, Winter 1994.

[105] Vajta, M., "New model reduction technique for a class of parabolic partial differential equations," Proc. of the IEEE International Conf. on Systems Engineering, Dayton OH, 1991, pp. 311–315.

[106] van Keulen, B., *$\mathcal{H}^\infty$-Control for Infinite Dimensional Systems: A State Space Approach*, Ph.D. Thesis, University of Groningen, 1993.

[107] van Keulen, B., and R. F. Curtain, "A state space approach to the mixed sensitivity minimization problem for

delay systems," *Proc. of Workshop on Control of Partial Differential Equations*, Italy, January 1993.

[108] Verma, M., and E. Jonckheere, "L^∞ Compensation with mixed sensitivity as a broadband matching problem," *Systems and Control Letters*, **4** (1984), pp. 125–129.

[109] Vidyasagar, M., *Control System Synthesis: A Factorization Approach*, MIT Press, Cambridge MA, 1985.

[110] Vidyasagar, M., and H. Kimura, "Robust controllers for uncertain linear multivariable systems," *Automatica*, **22** (1986), pp. 85–94.

[111] Willems, J. C., *The Analysis of Feedback Systems*, MIT Press, Cambridge MA, 1971.

[112] Wu, N. E., "Distributed parameter control system synthesis: an iterative approach" Proc. 1990 American Control Conf., San Diego CA, June 1990, pp. 1589–1595.

[113] Wu, N. E., and G. Gu, "Discrete Fourier transform and H^2 approximation," *IEEE Trans. Automatic Control*, **35** (1990), pp. 1044–1046.

[114] Wu, N. E., and E. B. Lee, "Feedback minimax synthesis for distributed systems," Proc. of the 27th IEEE Conf. on Decision and Control, Austin TX, December 1988, pp. 492–496.

[115] Yamamoto, Y., "Correspondence of internal and external stability: realization, transfer functions and complex analysis," in *Realization and Modelling in System Theory*, M. A. Kaashoek, J. H. van Schuppen, A. C. M. Ran Eds., Birkhäuser, Boston, 1990, pp. 61–72.

[116] Youla, D. C., H. A. Jabr, and J. J. Bongiorno Jr., "Modern Wiener Hopf design of optimal controllers: part II," *IEEE Transactions on Automatic Control*, **AC-21** (1976), pp. 319–338.

[117] Young, N. J., *Introduction to Hilbert Space*, Cambridge, 1988.

[118] Young, N. J., "An algorithm for the super-optimal sensitivity-minimising controller," *Proc. Workshop on New Perspectives in Industrial Control Design Using H_∞ Methods,* Oxford, 1986.

[119] Zames, G., "Feedback and optimal sensitivity: model reference transformations, multiplicative seminorms and approximate inverses," *IEEE Transactions on Automatic Control* **AC–26** (1981), pp. 301–320.

[120] Zames, G., and S. K. Mitter, "A note on essential spectrum and norms of mixed Hankel – Toeplitz operators," *Systems and Control Letters* **10** (1988), pp. 159–165.

[121] Zames, G., A. Tannenbaum, and C. Foias, "Optimal H^∞ interpolation: a new approach," Proceedings of the 25th IEEE Conf. on Decision and Control, Athens, Greece, December 1986, pp. 350–355.

[122] Zhou, K., and P. P. Khargonekar, "On the weighted sensitivity minimization problem for delay systems," *Systems and Control Letters*, **8** (1987), pp. 307–312.

List of Theorems, Lemmas, and Assumptions

Index

Lecture Notes in Control and Information Sciences

Edited by M. Thoma

1992–1995 Published Titles:

Vol. 167: Rao, Ming
Integrated System for Intelligent Control
133 pp. 1992 [3-540-54913-7]

Vol. 168: Dorato, Peter; Fortuna, Luigi; Muscato, Giovanni
Robust Control for Unstructured Perturbations: An Introduction
118 pp. 1992 [3-540-54920-X]

Vol. 169: Kuntzevich, Vsevolod M.; Lychak, Michael
Guaranteed Estimates, Adaptation and Robustness in Control Systems
209 pp. 1992 [3-540-54925-0]

Vol. 170: Skowronski, Janislaw M.; Flashner, Henryk; Guttalu, Ramesh S. (Eds)
Mechanics and Control. Proceedings of the 4th Workshop on Control Mechanics, January 21-23, 1991, University of Southern California, USA
302 pp. 1992 [3-540-54954-4]

Vol. 171: Stefanidis, P.; Paplinski, A.P.; Gibbard, M.J.
Numerical Operations with Polynomial Matrices: Application to Multi-Variable Dynamic Compensator Design
206 pp. 1992 [3-540-54992-7]

Vol. 172: Tolle, H.; Ersü, E.
Neurocontrol: Learning Control Systems Inspired by Neuronal Architectures and Human Problem Solving Strategies
220 pp. 1992 [3-540-55057-7]

Vol. 173: Krabs, W.
On Moment Theory and Controllability of Non-Dimensional Vibrating Systems and Heating Processes
174 pp. 1992 [3-540-55102-6]

Vol. 174: Beulens, A.J. (Ed.)
Optimization-Based Computer-Aided Modelling and Design. Proceedings of the First Working Conference of the New IFIP TC 7.6 Working Group, The Hague, The Netherlands, 1991
268 pp. 1992 [3-540-55135-2]

Vol. 175: Rogers, E.T.A.; Owens, D.H.
Stability Analysis for Linear Repetitive Processes
197 pp. 1992 [3-540-55264-2]

Vol. 176: Rozovskii, B.L.; Sowers, R.B. (Eds)
Stochastic Partial Differential Equations and their Applications. Proceedings of IFIP WG 7.1 International Conference, June 6-8, 1991, University of North Carolina at Charlotte, USA
251 pp. 1992 [3-540-55292-8]

Vol. 177: Karatzas, I.; Ocone, D. (Eds)
Applied Stochastic Analysis. Proceedings of a US-French Workshop, Rutgers University, New Brunswick, N.J., April 29-May 2, 1991
317 pp. 1992 [3-540-55296-0]

Vol. 178: Zolésio, J.P. (Ed.)
Boundary Control and Boundary Variation. Proceedings of IFIP WG 7.2 Conference, Sophia-Antipolis, France, October 15-17, 1990
392 pp. 1992 [3-540-55351-7]

Vol. 179: Jiang, Z.H.; Schaufelberger, W.
Block Pulse Functions and Their Applications in Control Systems
237 pp. 1992 [3-540-55369-X]

Vol. 180: Kall, P. (Ed.)
System Modelling and Optimization. Proceedings of the 15th IFIP Conference, Zurich, Switzerland, September 2-6, 1991
969 pp. 1992 [3-540-55577-3]

Vol. 181: Drane, C.R.
Positioning Systems - A Unified Approach
168 pp. 1992 [3-540-55850-0]

Vol. 182: Hagenauer, J. (Ed.)
Advanced Methods for Satellite and Deep Space Communications. Proceedings of an International Seminar Organized by Deutsche Forschungsanstalt für Luft-und Raumfahrt (DLR), Bonn, Germany, September 1992
196 pp. 1992 [3-540-55851-9]

Vol. 183: Hosoe, S. (Ed.)
Robust Control. Proceesings of a Workshop held in Tokyo, Japan, June 23-24, 1991
225 pp. 1992 [3-540-55961-2]

Vol. 184: Duncan, T.E.; Pasik-Duncan, B. (Eds)
Stochastic Theory and Adaptive Control. Proceedings of a Workshop held in Lawrence, Kansas, September 26-28, 1991
500 pp. 1992 [3-540-55962-0]

Vol. 185: Curtain, R.F. (Ed.); Bensoussan, A.; Lions, J.L.(Honorary Eds)
Analysis and Optimization of Systems: State and Frequency Domain Approaches for Infinite-Dimensional Systems. Proceedings of the 10th International Conference, Sophia-Antipolis, France, June 9-12, 1992.
648 pp. 1993 [3-540-56155-2]

Vol. 186: Sreenath, N.
Systems Representation of Global Climate Change Models. Foundation for a Systems Science Approach.
288 pp. 1993 [3-540-19824-5]

Vol. 187: Morecki, A.; Bianchi, G.; Jaworeck, K. (Eds)
RoManSy 9: Proceedings of the Ninth CISM-IFToMM Symposium on Theory and Practice of Robots and Manipulators.
476 pp. 1993 [3-540-19834-2]

Vol. 188: Naidu, D. Subbaram
Aeroassisted Orbital Transfer: Guidance and Control Strategies
192 pp. 1993 [3-540-19819-9]

Vol. 189: Ilchmann, A.
Non-Identifier-Based High-Gain Adaptive Control
220 pp. 1993 [3-540-19845-8]

Vol. 190: Chatila, R.; Hirzinger, G. (Eds)
Experimental Robotics II: The 2nd International Symposium, Toulouse, France, June 25-27 1991
580 pp. 1993 [3-540-19851-2]

Vol. 191: Blondel, V.
Simultaneous Stabilization of Linear Systems
212 pp. 1993 [3-540-19862-8]

Vol. 192: Smith, R.S.; Dahleh, M. (Eds)
The Modeling of Uncertainty in Control Systems
412 pp. 1993 [3-540-19870-9]

Vol. 193: Zinober, A.S.I. (Ed.)
Variable Structure and Lyapunov Control
428 pp. 1993 [3-540-19869-5]

Vol. 194: Cao, Xi-Ren
Realization Probabilities: The Dynamics of Queuing Systems
336 pp. 1993 [3-540-19872-5]

Vol. 195: Liu, D.; Michel, A.N.
Dynamical Systems with Saturation Nonlinearities: Analysis and Design
212 pp. 1994 [3-540-19888-1]

Vol. 196: Battilotti, S.
Noninteracting Control with Stability for Nonlinear Systems
196 pp. 1994 [3-540-19891-1]

Vol. 197: Henry, J.; Yvon, J.P. (Eds)
System Modelling and Optimization
975 pp approx. 1994 [3-540-19893-8]

Vol. 198: Winter, H.; Nüßer, H.-G. (Eds)
Advanced Technologies for Air Traffic Flow Management
225 pp approx. 1994 [3-540-19895-4]

Vol. 199: Cohen, G.; Quadrat, J.-P. (Eds)
11th International Conference on Analysis and Optimization of Systems – Discrete Event Systems: Sophia-Antipolis, June 15–16–17, 1994
648 pp. 1994 [3-540-19896-2]

Vol. 200: Yoshikawa, T.; Miyazaki, F. (Eds)
Experimental Robotics III: The 3rd International Symposium, Kyoto, Japan, October 28-30, 1993
624 pp. 1994 [3-540-19905-5]

Vol. 201: Kogan, J.
Robust Stability and Convexity
192 pp. 1994 [3-540-19919-5]

Vol. 202: Francis, B.A.; Tannenbaum, A.R. (Eds)
Feedback Control, Nonlinear Systems, and Complexity
288 pp. 1995 [3-540-19943-8]

Vol. 203: Popkov, Y.S.
Macrosystems Theory and its Applications: Equilibrium Models
344 pp. 1995 [3-540-19955-1]

Vol. 204: Takahashi, S.; Takahara, Y.
Logical Approach to Systems Theory
192 pp. 1995 [3-540-19956-X]

Vol. 205: Kotta, U.
Inversion Method in the Discrete-time Nonlinear Control Systems Synthesis Problems
168 pp. 1995 [3-540-19966-7]

Vol. 206: Aganovic, Z.;.Gajic, Z.
Linear Optimal Control of Bilinear Systems with Applications to Singular Perturbations and Weak Coupling
133 pp. 1995 [3-540-19976-4]

Vol. 207: Gabasov, R.; Kirillova, F.M.; Prischepova, S.V. (Eds)
Optimal Feedback Control
224 pp. 1995 [3-540-19991-8]

Vol. 208: Khalil, H.K.; Chow, J.H.; Ioannou, P.A. (Eds)
Proceedings of Workshop on Advances in Control and its Applications
348 pp. 1995 [3-540-19993-4]